National Water-Quality Assessment Program

Use of Classes Based on Redox and Groundwater Age to Characterize the Susceptibility of Principal Aquifers to Changes in Nitrate Concentrations, 1991 to 2010

Scientific Investigations Report 2012–5220

U.S. Department of the Interior
U.S. Geological Survey

Use of Classes Based on Redox and Groundwater Age to Characterize the Susceptibility of Principal Aquifers to Changes in Nitrate Concentrations, 1991 to 2010

By P.B. McMahon

National Water-Quality Assessment Program

Scientific Investigations Report 2012–5220

U.S. Department of the Interior
U.S. Geological Survey

U.S. Department of the Interior
KEN SALAZAR, Secretary

U.S. Geological Survey
Marcia K. McNutt, Director

U.S. Geological Survey, Reston, Virginia: 2012

For more information on the USGS—the Federal source for science about the Earth, its natural and living resources, natural hazards, and the environment, visit http://www.usgs.gov or call 1–888–ASK–USGS.
For an overview of USGS information products, including maps, imagery, and publications,
visit http://www.usgs.gov/pubprod

To order this and other USGS information products, visit http://store.usgs.gov

Suggested citation:
McMahon, P.B., 2012, Use of classes based on redox and groundwater age to characterize the susceptibility of principal aquifers to changes in nitrate concentrations, 1991 to 2010: U.S. Geological Survey Scientific Investigations Report 2012–5220, 41 p.

Foreword

The U.S. Geological Survey (USGS) is committed to providing the Nation with reliable scientific information that helps to enhance and protect the overall quality of life and that facilitates effective management of water, biological, energy, and mineral resources (*http://www.usgs.gov/*). Information on the Nation's water resources is critical to ensuring long-term availability of water that is safe for drinking and recreation and is suitable for industry, irrigation, and fish and wildlife. Population growth and increasing demands for water make the availability of that water, measured in terms of quantity and quality, even more essential to the long-term sustainability of our communities and ecosystems.

The USGS implemented the National Water-Quality Assessment (NAWQA) Program in 1991 to support national, regional, State, and local information needs and decisions related to water-quality management and policy (*http://water.usgs.gov/nawqa*). The NAWQA Program is designed to answer: What is the quality of our Nation's streams and groundwater? How are conditions changing over time? How do natural features and human activities affect the quality of streams and groundwater, and where are those effects most pronounced? By combining information on water chemistry, physical characteristics, stream habitat, and aquatic life, the NAWQA Program aims to provide science-based insights for current and emerging water issues and priorities. From 1991 to 2001, the NAWQA Program completed interdisciplinary assessments and established a baseline understanding of water-quality conditions in 51 of the Nation's river basins and aquifers, referred to as Study Units (*http://water.usgs.gov/nawqa/studies/study_units.html*).

In the second decade of the Program (2001–2012), a major focus is on regional assessments of water-quality conditions and trends. These regional assessments are based on major river basins and principal aquifers, which encompass larger regions of the country than the Study Units. Regional assessments extend the findings in the Study Units by filling critical gaps in characterizing the quality of surface water and groundwater, and by determining water-quality status and trends at sites that have been consistently monitored for more than a decade. In addition, the regional assessments continue to build an understanding of how natural features and human activities affect water quality. Many of the regional assessments employ modeling and other scientific tools, developed on the basis of data collected at individual sites, to help extend knowledge of water quality to unmonitored, yet comparable areas within the regions. The models thereby enhance the value of our existing data and our understanding of the hydrologic system. In addition, the models are useful in evaluating various resource-management scenarios and in predicting how our actions, such as reducing or managing nonpoint and point sources of contamination, land conversion, and altering flow and (or) pumping regimes, are likely to affect water conditions within a region.

Other activities planned during the second decade include continuing national syntheses of information on pesticides, volatile organic compounds (VOCs), nutrients, trace elements, and aquatic ecology; and continuing national topical studies on the fate of agricultural chemicals, effects of urbanization on stream ecosystems, bioaccumulation of mercury in stream ecosystems, effects of nutrient enrichment on stream ecosystems, and transport of contaminants to public-supply wells.

The USGS aims to disseminate credible, timely, and relevant science information to address practical and effective water-resource management and strategies that protect and restore water quality. We hope this NAWQA publication will provide you with insights and information to meet your needs, and will foster increased citizen awareness and involvement in the protection and restoration of our Nation's waters.

The USGS recognizes that a national assessment by a single program cannot address all water-resource issues of interest. External coordination at all levels is critical for cost-effective management, regulation, and conservation of our Nation's water resources. The NAWQA Program, therefore, depends on advice and information from other agencies—Federal, State, regional, interstate, Tribal, and local—as well as nongovernmental organizations, industry, academia, and other stakeholder groups. Your assistance and suggestions are greatly appreciated.

William H. Werkheiser
USGS Associate Director for Water

Contents

Figures

Appendix Figure

Tables

Appendix Table

Conversion Factors

Multiply	By	To obtain
	Length	
centimeter (cm)	0.3937	inch (in.)
millimeter (mm)	0.03937	inch (in.)
meter (m)	3.281	foot (ft)
kilometer (km)	0.6214	mile (mi)
	Volume	
liter (L)	33.82	ounce, fluid (fl. oz)
liter (L)	0.2642	gallon (gal)
Mass	Mass	
gram (g)	0.03527	ounce, avoirdupois (oz)

Concentrations of chemical constituents in water are given either in milligrams per liter (mg/L) or micrograms per liter (µg/L). Concentrations of tritium are given in tritium units (TU).

Use of Classes Based on Redox and Groundwater Age to Characterize the Susceptibility of Principal Aquifers to Changes in Nitrate Concentrations, 1991 to 2010

By P.B. McMahon

Abstract

The National Water-Quality Assessment (NAWQA) Program of the U.S. Geological Survey is using multiple approaches to measure and explain trends in concentrations of nitrate in principal aquifers of the United States. Near decadal sampling of selected well networks is providing information on where long-term changes in nitrate concentrations have occurred. Because those studies do not include all the NAWQA well networks, a determination has yet to be made as to what might be expected in networks from which time-series data have not been collected. Characterizing aquifer susceptibility to changes in nitrate concentrations on the basis of data collected from all the NAWQA well networks would be a step toward extrapolating findings from those studies to broader regions.

In this study, water samples collected from 6,593 wells in 39 principal aquifers and 5 alluvial aquifers (collected from 1991 to 2010) were assigned to four redox-age classes on the basis of concentrations of dissolved oxygen and various indicators of groundwater age. The redox-age assignments were then used to characterize the susceptibility of principal aquifers to changes in nitrate concentrations. Aquifer areas (as defined by well networks) in which at least 75 percent of the samples were classified as oxic-potentially young were considered to have a high susceptibility to changes in nitrate concentrations. Aquifer areas were considered to have a medium susceptibility if at least 25 percent and less than 75 percent of the samples were classified as oxic-potentially young. Aquifer areas were considered to have a low susceptibility if less than 25 percent of the samples were classified as oxic-potentially young.

The three primary well types sampled by NAWQA (shallow monitoring wells near the water table, domestic wells, and public-supply wells) generally represent different depth zones and (or) areas of the principal aquifers. For the parts of aquifers near the water table in agricultural areas, the aquifers most susceptible to changes in nitrate concentrations were the Columbia Plateau basin-fill aquifers, Eastern glacial aquifers, and the West-central glacial aquifers. None of the aquifers had a low susceptibility to changes in nitrate concentrations. For the parts of aquifers that provide domestic water supplies, the aquifers most susceptible to changes in nitrate concentrations were the Northern Atlantic Coastal Plain aquifer system and the Early Mesozoic Basin, Valley and Ridge carbonate-rock, and Piedmont and Blue Ridge crystalline-rock aquifers in the eastern United States; the Ozark Plateaus aquifer system in parts of Missouri and Arkansas; and the Central Valley, Columbia Plateau basaltic-rock, and Snake River Plain basaltic-rock aquifer systems in the West. The least susceptible aquifers were the Texas Coastal Uplands and Denver Basin aquifer systems. For the parts of aquifers that provide public water supplies, the aquifers most susceptible to changes in nitrate concentrations were the Eastern glacial aquifers and the California Coastal Basin, Basin and Range basin-fill, and High Plains aquifers in the West. The least susceptible aquifer was the Cambrian-Ordovician aquifer system in the upper Midwest.

Principal-aquifer lithology groups with the largest percentage of domestic-well networks considered to have a high susceptibility to changes in nitrate concentrations were the basaltic- and other volcanic-rock aquifer systems, carbonate-rock aquifers, and crystalline-rock aquifers. The lithology groups with the smallest percentage of networks considered to have a high susceptibility to changes in nitrate concentrations were the glacial aquifers and sandstone aquifers. There are important geologic differences between the aquifer lithology groups with high and low susceptibilities to changes in nitrate concentrations. The relatively large percentage of high-susceptibililty networks in the basaltic- and other volcanic-rock aquifer systems, carbonate-rock aquifers, and crystalline-rock aquifers may indicate the importance of fractures and karst features in promoting the rapid movement of oxic-potentially young groundwater in those aquifers. The relatively small percentage of high-susceptibility networks in the glacial and sandstone aquifers reflects geologic characteristics of those aquifers that support anoxic redox conditions (high electron-donor content) and inhibit water movement (fine-grained confining layers).

For networks of monitoring and domestic wells that were approximately collocated, the monitoring-well networks had the higher percentage of samples classified as oxic-potentially young, indicating that susceptibility tended to be higher at the shallower depths of the monitoring wells. For networks of domestic and public-supply wells that were approximately collocated, the public-supply wells had the higher percentage of samples classified as oxic-potentially young, indicating that susceptibility tended to be higher in the vicinity of public-supply wells than in the vicinity of domestic wells even though the public-supply wells had larger median well depths. Previous studies found that high rates of pumping in public-supply wells with long screens induced more rapid downward movement of young groundwater than did domestic wells, which had shorter screens and were less heavily pumped. The data from this study are generally consistent with those findings.

Introduction

Is groundwater quality getting better or worse, why, and what will happen in the future? These are some of the important questions being addressed by the National Water-Quality Assessment (NAWQA) Program of the U.S. Geological Survey (USGS). The NAWQA Program is using multiple approaches to measure and explain trends in concentrations of nitrate in principal aquifers of the United States. Near decadal sampling of selected NAWQA well networks has provided a preliminary determination of where changes in concentrations of chloride, dissolved solids, and nitrate in groundwater have occurred (Rupert, 2008; Lindsey and Rupert, 2012). Because those studies do not include all the NAWQA well networks, a determination has yet to be made as to what might be expected in networks from which time-series data have not been collected. Characterizing aquifer susceptibility to changes in nitrate concentrations on the basis of data collected from all the NAWQA well networks would be a step toward extrapolating findings from those studies to broader regions.

Long-term changes in concentrations of nitrate in groundwater are controlled by factors such as nitrogen input history at the land surface, denitrification in the aquifer, and location in the flow system (Clark and others, 2008; Burow and others, 2008a,b; Kauffman and others, 2001; McMahon and others, 2008a,b). Nitrate concentrations in groundwater increased over several decades in many agricultural areas of the United States following the dramatic increase in fertilizer usage that began in the late 1940s (see summary by Puckett and others, 2011). Denitrification is the microbial reduction of nitrate to nitrogen gas (N_2) and in some aquifers it is an important process for decreasing nitrate concentrations (Böhlke and others, 2002; Green and others, 2008; Tesoriero and Puckett, 2011). Aquifers containing oxic shallow groundwater are more susceptible to changes in nitrate concentrations than aquifers containing anoxic deep groundwater (Dubrovsky and others, 2010). This pattern occurs because nitrate persists in oxic groundwater and is removed by denitrification in

anoxic groundwater, and deep groundwater generally is older, sometimes predating nitrogen inputs by humans at the land surface, and contains a broader mix of water of differing ages and sources than shallow groundwater. Thus, redox-age classifications could provide a framework for characterizing the susceptibility of aquifers to changes in nitrate concentrations and such a framework could be used for interpreting time-series monitoring data.

To a certain extent, this type of redox-age assessment has already been done using statistically based models of groundwater vulnerability (Rupert, 1998; Nolan and Hitt, 2006; Gurdak and Qi, 2006; Rupert and Plummer, 2009). Vulnerability models sometimes incorporate redox and groundwater-age information, but use surrogate variables, such as soil type and well depth (Nolan and Hitt, 2006). More direct measures of redox conditions and groundwater age are available, such as concentrations of dissolved oxygen, chlorofluorocarbons, and sulfur-hexafluoride, and detections of tritium, pesticide compounds, or volatile organic compounds (VOCs).

The purpose of this report is to characterize the susceptibility of selected principal aquifers of the United States to changes in nitrate concentrations on a basis of the redox-age classification scheme developed in this report. Redox classes are defined by concentrations of dissolved oxygen. Groundwater-age classes are defined by concentrations of tritium, nitrate, pesticide compounds, VOCs, chlorofluorocarbons, sulfur hexafluoride, and (or) helium. The redox-age classification scheme uses NAWQA water-quality data collected from 39 principal aquifers and 5 alluvial aquifers from 1991 to 2010.

Methods

This section describes the well networks and water-quality data used in the study, and the principal aquifers in which the networks are located. In addition, the redox and groundwater-age classes used to characterize the aquifers are defined.

Well Selection

Wells included in this study were used in NAWQA studies designed to describe the quality of water withdrawn from principal aquifers and used for drinking (termed major-aquifer studies or MASs, and source-water studies or DWGSs), and studies of shallow groundwater within specific land-use settings (termed land-use studies or LUSs). MASs focused on the quality of groundwater resources without being linked to a specific land use and used data mostly from existing domestic wells. DWGSs focused on the quality of groundwater from public-supply wells. Water samples from MAS and DWGS wells were collected before any treatment or pressure tanks and therefore do not represent water consumed for drinking. LUSs targeted the uppermost recently recharged groundwater to identify the effects of the overlying land use and used data mostly from monitoring wells and some production wells. Generally, MAS networks covered larger geographic areas and their wells were deeper than LUSs. Gilliom and others (1995) presented a general discussion of NAWQA well networks. Individual MASs, DWGSs, and LUSs are described in reports for individual NAWQA study areas (U.S. Geological Survey, 2011). Data for a total of 6,593 wells from networks in 39 principal aquifers and 5 alluvial aquifers were used in this study. Locations of the principal aquifers are shown in fig. 1.

Sources of Data

Many of the NAWQA wells have been sampled more than once, but the water-quality data used in this study primarily represent the most recently collected water sample from each well. Water samples included in this study were collected from 1991 to 2010. Methods for collecting and analyzing groundwater samples for the NAWQA program are well documented (U.S. Geological Survey, 2011) and are not repeated here. Water-quality and groundwater-age data of primary interest in this study are concentrations of dissolved oxygen, nitrate, pesticide compounds, volatile organic compounds (VOCs), chlorofluorocarbons, sulfur hexafluoride, tritium, and helium. Dissolved oxygen was measured in the field at the time of sample collection. Nitrate, pesticide compounds, and VOCs were analyzed at the USGS National Water Quality Laboratory in Denver, Colorado. Chlorofluorocarbons and sulfur hexafluoride were analyzed at the USGS Chlorofluorocarbon Laboratory in Reston, Virginia. Tritium was measured at the USGS Tritium Laboratory in Menlo Park, California, the USGS Noble Gas Laboratory in Denver, Colorado, or the Noble Gas Laboratory of the Lamont-Doherty Earth Observatory in Palisades, New York. Helium was measured at the USGS Noble Gas Laboratory in Denver, Colorado, or the Noble Gas Laboratory of the Lamont-Doherty Earth Observatory in Palisades, New York. The data can be found in the USGS National Water Information System (NWIS) or in Hinkle and others (2010).

Aquifer Groups and Geology

This study primarily examined 39 principal aquifers (fig. 1). Principal aquifers are defined as aquifers that are regionally extensive and can yield useable quantities of water (U.S. Geological Survey, 2003). The aquifers are broadly grouped into eight lithologic groups: basaltic and other volcanic rocks, carbonate rocks, crystalline rocks, glacial sand and gravel, sandstone and carbonate rocks, sandstone, semiconsolidated sand, and unconsolidated sand and gravel (U.S. Geological Survey, 2003). Five relatively small alluvial aquifers also were examined and they are considered to be part of the unconsolidated sand and gravel lithology group.

Redox Classification

Redox conditions in many of the principal aquifers were described on a regional basis (McMahon and others, 2009) using the redox framework developed by McMahon and Chapelle (2008). That framework uses a dissolved-oxygen concentration of 0.5 milligram per liter (mg/L) as the threshold between oxic and anoxic conditions. The framework has additional redox subclasses for anoxic conditions, but for the purposes of this report only two redox classes are considered—oxic (dissolved oxygen concentration greater than or equal to 0.5 mg/L) and anoxic (dissolved oxygen concentration less than 0.5 mg/L). Consideration of just two redox classes is appropriate for this report because oxygen reduction typically is the first redox process to occur in groundwater and denitrification typically is the first anoxic redox process to follow oxygen reduction when nitrate is present in groundwater (Chapelle and others, 1995; McMahon and Chapelle, 2008). The susceptibility of aquifers to changes in nitrate concentrations would be greater in oxic groundwater than in anoxic groundwater (Rupert, 2008), although the susceptibility also could be high in anoxic groundwater that is actively undergoing denitrification.

The use of a dissolved-oxygen concentration of 0.5 mg/L as the threshold for onset of denitrification probably is conservative. As discussed by Green and others (2010), mixing in heterogeneous aquifers and in well screens can result in the co-occurrence of geochemical indicators of denitrification with dissolved oxygen concentrations greater than 0.5 mg/L. Several field studies have reported apparent threshold concentrations for the onset of denitrification in the range of about 1 to 2 mg/L (Böhlke and others, 2002; McMahon and others, 2004; Böhlke and others, 2007; Green and others, 2008; Tesoriero and Puckett, 2011).

Groundwater-Age Classification

The substantial increase in fertilizer usage in the United States beginning in about the late 1940s is an important event in the context of this study because fertilizer represents the largest single anthropogenic source of nitrogen in the country (Dubrovsky and others, 2010), and an increase in usage of fertilizer has been linked to increased concentrations of nitrate in groundwater (see review by Puckett and others, 2011). Given this history, a useful tracer of groundwater age for this study would differentiate between water recharged before and after the early 1950s. For the purposes of this report these waters are referred to as old and young groundwater, respectively. The susceptibility of aquifers to changes in nitrate concentrations would be greater in young groundwater than in old groundwater.

Tritium was used to differentiate between old and young groundwater in this study. Tritium is a radioactive isotope of hydrogen with a half-life of 12.32 years (Lucas

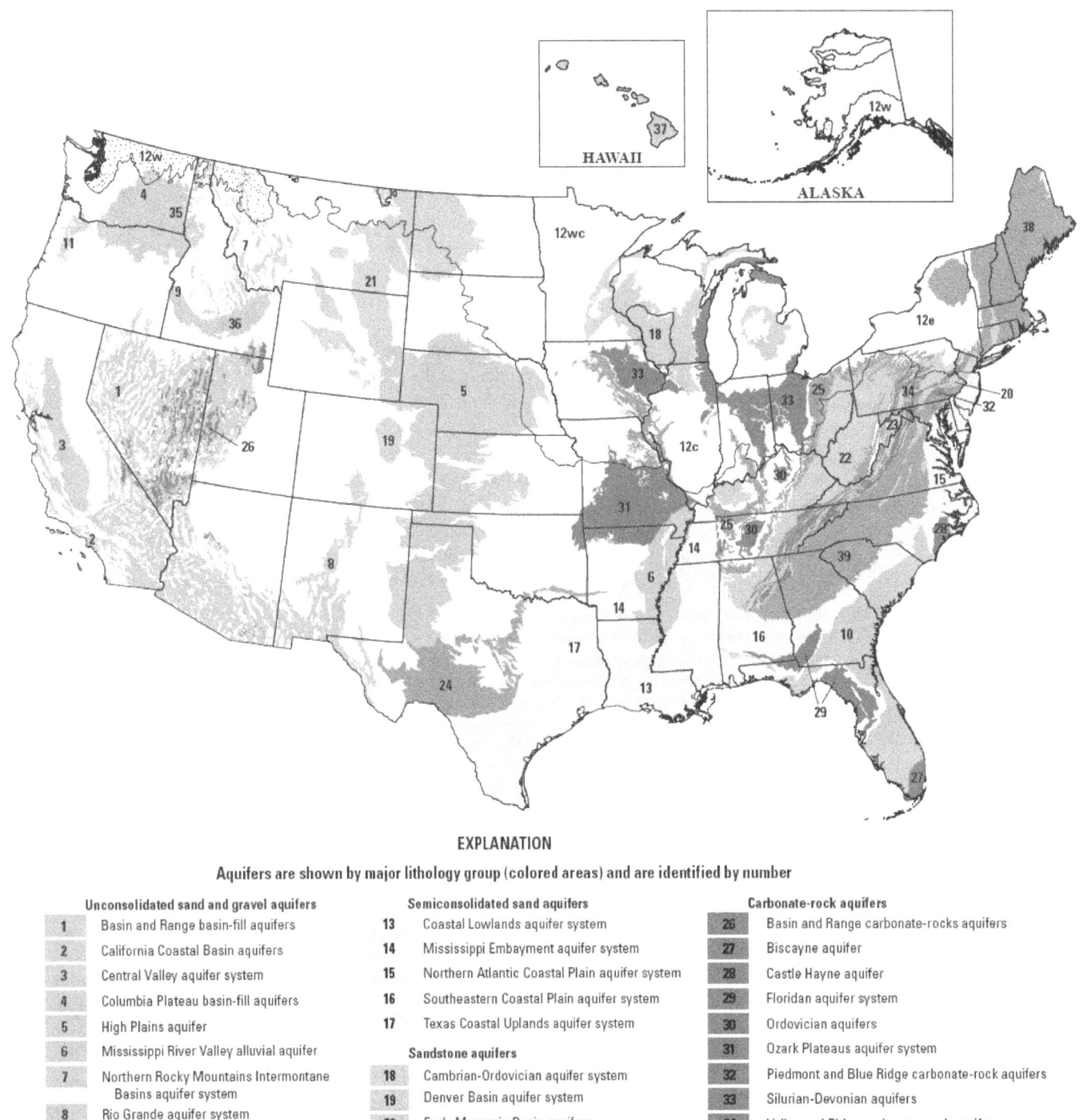

EXPLANATION

Aquifers are shown by major lithology group (colored areas) and are identified by number

Unconsolidated sand and gravel aquifers

1	Basin and Range basin-fill aquifers
2	California Coastal Basin aquifers
3	Central Valley aquifer system
4	Columbia Plateau basin-fill aquifers
5	High Plains aquifer
6	Mississippi River Valley alluvial aquifer
7	Northern Rocky Mountains Intermontane Basins aquifer system
8	Rio Grande aquifer system
9	Snake River Plain basin-fill aquifers
10	Surficial aquifer system
11	Willamette Lowland aquifer system

Glacial sand and gravel aquifers

12e	Eastern glacial aquifers
12c	Central glacial aquifers
12wc	West-central glacial aquifers
12w	Western glacial aquifers

Semiconsolidated sand aquifers

13	Coastal Lowlands aquifer system
14	Mississippi Embayment aquifer system
15	Northern Atlantic Coastal Plain aquifer system
16	Southeastern Coastal Plain aquifer system
17	Texas Coastal Uplands aquifer system

Sandstone aquifers

18	Cambrian-Ordovician aquifer system
19	Denver Basin aquifer system
20	Early Mesozoic Basin aquifers
21	Lower Tertiary aquifers
22	Pennsylvanian aquifers
23	Valley and Ridge clastic-rock aquifers

Sandstone and carbonate-rock aquifers

24	Edwards-Trinity aquifer system
25	Mississippian aquifers

Carbonate-rock aquifers

26	Basin and Range carbonate-rocks aquifers
27	Biscayne aquifer
28	Castle Hayne aquifer
29	Floridan aquifer system
30	Ordovician aquifers
31	Ozark Plateaus aquifer system
32	Piedmont and Blue Ridge carbonate-rock aquifers
33	Silurian-Devonian aquifers
34	Valley and Ridge carbonate-rock aquifers

Basaltic- and other volcanic-rock aquifers

35	Columbia Plateau basaltic-rock aquifer system
36	Snake River Plain basaltic-rock aquifer system
37	Hawaiian volcanic-rock aquifers

Crystalline-rock aquifers

38	New York and New England crystalline-rock aquifers (unofficial name)
39	Piedmont and Blue Ridge crystalline-rock aquifers

Figure 1.   Location of selected principal aquifers in the United States (modified from U.S. Geological Survey, 2003).

and Unterweger, 2000). Small concentrations of tritium are produced naturally by interactions between the atmosphere and cosmic rays. It is an excellent tracer of water movement because it is part of the water molecule. In general, tritium in groundwater originates from precipitation. Because tritium is radioactive, its concentration in groundwater decreases over time as a result of radioactive decay. Before the onset of atmospheric testing of nuclear weapons in about 1953 (prebomb), the tritium content of precipitation in the conterminous United States probably ranged from about 2 to 8 tritium units (TU) (Kaufman and Libby, 1954; Thatcher, 1962). As a result of radioactive decay, groundwater derived from precipitation that fell before 1953 would have contained less than 0.5 TU tritium in 2010 (the most recent samples used in this study), but it could have contained upwards of about 1 TU in 1991 (the earliest samples used in this study). The tritium content of precipitation increased substantially after the onset of atmospheric nuclear weapons testing but has slowly decreased from its peak in the early 1960s. Even with the variability in tritium content of precipitation over time, most groundwater in the United States exclusively derived from precipitation that fell since 1953 (postbomb) contained more than 0.5 TU in 2010. On the basis of this information, water samples with tritium concentrations less than 0.5 TU were considered to be potentially old groundwater (recharged before the early 1950s), and water samples with tritium concentrations greater than or equal to 0.5 TU were considered to be potentially young groundwater (recharged after the early 1950s). For comparison, several other studies have used tritium concentrations ranging from about 0.2 to 1 TU as the cutoff between old and young groundwater (Michel and Schroeder, 1994; Plummer and others, 2004; Manning and others, 2005; Landon and others, 2010a).

Because only 39 percent of the samples that were assigned to a groundwater-age class had tritium data, other indicators of groundwater age were used for the samples that did not have tritium data. These indicators include detections of pesticide compounds and (or) VOCs; elevated concentrations of nitrate; and (or) dating with chloroflurocarbons, sulfur hexafluoride, or tritium/helium-3 (Plummer and others, 1993; Kolpin and others, 1995; Busenberg and Plummer, 2000; Shelton and others, 2001; Manning and others, 2005; Plummer and others, 2008). The water-quality data set used in this study contained information for as many as 155 pesticide compounds and 85 VOCs. Minimum detection levels for the pesticide compounds ranged from 0.000057 to 0.021 microgram per liter (μg/L). For the VOCs, minimum detection levels ranged from 0.001 to 0.3 μg/L. Samples that did not have tritium data but had a detection of a pesticide compound or a VOC were considered to be potentially young groundwater. Samples that did not have tritium data but had a nitrate concentration greater than 1.3 milligrams of nitrogen per

liter (mg-N/L) also were considered to be potentially young groundwater. This concentration represents the 75th percentile concentration of nitrate in groundwater samples with tritium concentrations less than 0.5 TU. For comparison, Nolan and Hitt (2003) proposed a national background nitrate concentration of about 1 mg-N/L and Mueller and Helsel (1996) proposed a background concentration of 2 mg-N/L. Background nitrate concentrations in groundwater are likely to vary regionally and locally, but that variability was not taken into account in this study. Some samples were dated using chloroflurocarbons, sulfur hexafluoride, or tritium/helium-3 (Hinkle and others, 2010), and those data were used to determine the presence of young groundwater in samples for which tritium data were unavailable. For samples that had no tritium, chlorofluorocarbon, sulfur-hexafluoride, and tritium/helium-3 data, or detections of a pesticide compound or VOC, and had a nitrate concentration of less than or equal to 1.3 mg-N/L, the age was considered to be potentially old. The criteria used to assign redox-age classes to groundwater samples are summarized in table 1.

The approach for classifying groundwater ages does not consider mixing, which is why the age determinations are qualified as being potentially old or young. It is likely that some groundwater classified as being potentially old contained a component of young groundwater, and vice versa (Weissmann and others, 2002; Manning and others, 2005; Plummer and others, 2008). For example, 54 percent of the samples that were classified as potentially old on the basis of tritium concentrations less than 0.5 TU would have been classified as potentially young using just the pesticide compound, VOC, and nitrate data. Twenty-one percent of the samples that were classified as being potentially young on the basis of tritium concentrations greater than or equal to 0.5 TU would have been classified as potentially old using just the pesticide compound, VOC, and nitrate data. Thus, some samples overlapped the age classes used in this report. Another potential limitation is that some VOCs could be present in groundwater recharged before 1950 either from natural sources or solvent and fuel use in the early 20th century. Chloroform, the most commonly detected VOC in the Nation's groundwater (Zogorski and others, 2006), has both natural and man-made sources (McCulloch, 2003). Although the presence of tritium, pesticide compounds, VOCs, or elevated concentrations of nitrate in a sample generally indicates that the sample contained a fraction of young groundwater (Plummer and others, 2008), it does not indicate how much. Techniques are available for estimating the fractions of old and young groundwater in samples (Plummer and others, 2003; Manning and others, 2005), but the data required for that analysis were not available for most of the samples used in this study. Ideally, one would analyze all the water samples for a comparable set of tracers that characterize groundwater

Table 1.   Criteria used to assign redox-age classes to groundwater samples. Young groundwater is defined as water recharged since the early 1950s, and old groundwater is defined as water recharged before the early 1950s. Age classes are labeled as potentially young or old because some samples probably represent a mixture of ages and the fractions of young and old water in them are unknown.

[≥, greater than or equal to; >, greater than; ≤, less than or equal to; < less than; mg/L, milligrams per liter; mg-N/L, milligrams of nitrogen per liter; TU, tritium units]

Class	Criteria
	Redox class
Oxic	Dissolved oxygen ≥0.5 mg/L
Anoxic	Dissolved oxygen <0.5 mg/L
	Age class
Potentially young	(a) If tritium data are available Tritium concentration ≥0.5 TU (b) If no tritium data are available Detection of at least one pesticide compound or Detection of at least one volatile organic compound or Nitrate concentration >1.3 mg-N/L or Dated using chlorofluorocarbons, sulfur hexafluoride, or tritium/helium-3
Potentially old	(c) If tritium data are available Tritium concentration <0.5 TU (d) If no tritium, chlorofluorocarbon, sulfur hexafluoride, or tritium/helium-3 data are available No detection of pesticide compounds and No detection of volatile organic compounds and Nitrate concentration ≤1.3 mg-N/L

age at multiple time scales to distinguish between water that is completely old or young, or is mixed (Landon and others, 2010a). Despite the limitations of the approach used in this report to classify groundwater age, the approach still provides useful information at the regional scale examined in this report, as is described in the next section.

Evaluation of Redox-Age Classes

Two approaches were used to evaluate whether the redox-age classes described in table 1 could provide useful information on the susceptibility of principal aquifers to changes in nitrate concentrations. In the first approach, redox-age classes assigned to water samples from as many as 6,489 wells were evaluated in relation to well depth, well type, and aquifer confinement to see if the redox-age classes made sense hydrologically. In the second approach, redox-age classes assigned to water samples from 1,111 trend wells that were sampled at near decadal time scales (Lindsey and Rupert, 2012) were evaluated in relation to changes in nitrate concentrations in those samples to determine which redox-age classes had the largest and smallest changes in nitrate concentrations.

Oxic-potentially young water was mostly associated with relatively shallow monitoring and domestic wells completed in unconfined aquifers, whereas anoxic-potentially old water was mostly associated with deeper domestic wells completed in unconfined and confined aquifers (table 2). In general, the median well depths associated with each of the four primary

redox-age classes increased in the order of anoxic-potentially young, oxic-potentially young, oxic-potentially old, and anoxic-potentially old (table 2). The fact that potentially young water came from shallower wells than potentially old water makes sense hydrologically and is consistent with what is known about groundwater-age stratigraphy in the principal aquifers (McMahon and others, 2011; Puckett and others, 2011).

Within a given redox class, the difference in median well depths between age criteria (a) and (b) was smaller than the difference between criteria (c) and (d) (tables 1 and 2). This could mean that age criteria (a) and (b), used to classify potentially young water, are more comparable than age criteria (c) and (d), used to classify potentially old water. For potentially old water, the median well depth was 1.6 to 2.0 times greater for samples classified using criterion (c) than it was for samples classified using criterion (d). This could indicate that small nitrate concentrations and the absence of detections of pesticide compounds and VOCs (criterion (d)) is not always indicative of old water. In some instances, the smaller median well depth for samples classified using criterion (d) might indicate young groundwater that was not impacted by anthropogenic chemicals, in which case criterion (d) would overestimate the amount of potentially old water in an aquifer. For the purpose of characterizing the susceptibility of an aquifer to changes in nitrate concentrations, overestimation of the amount of potentially old water is probably less of an issue in anoxic water than oxic water because of the higher denitrification potential in

Table 2. Redox-age classes for water samples collected from principal aquifers in the United States and median well depth, well type, and aquifer confinement.

[Refer to table 1 for the definition of age criteria a through d; m, meters; %, percentage of samples]

Redox-age class	Well depth below land surface		Well type				Aquifer confinement		
	Number of samples	Median (m)	Number of samples	Shallow monitoring wells (%)	Domestic wells (%)	Public-supply wells (%)	Number of samples	Confined (%)	Unconfined (%)
Oxic-potentially young, criterion (a)	1,489	31	1,393	46	43	10	1,106	18.1	81.9
Oxic-potentially young, criterion (b)	2,540	22	2,380	46	41	13	1,720	12.5	87.5
Oxic-potentially old, criterion (c)	321	70	301	16	66	18	254	20.5	79.5
Oxic-potentially old, criterion (d)	319	43	305	25	55	20	181	21.5	78.5
Anoxic-potentially young, criterion (a)	470	18	428	49	43	8	312	24.4	75.6
Anoxic-potentially young, criterion (b)	761	19	732	45	42	14	493	26.8	73.2
Anoxic-potentially old, criterion (c)	267	93	249	12	61	27	202	77.7	22.3
Anoxic-potentially old, criterion (d)	322	46	313	24	55	20	197	33.5	66.5

anoxic water. For each redox class, the median well depth for potentially old water classified using criterion (d) still was larger than the median well depth for potentially young water, regardless of whether age criterion (a) or (b) was used to classify that water. Overall, age criterion (d) probably is indicative of old water in some instances and young water in others; of the eight possible redox-age classes (tables 1 and 2), the oxic-potentially old classification based on age criterion (d) probably has the greatest uncertainty with respect to characterizing aquifer susceptibility to changes in nitrate concentrations. The oxic-potentially old, criterion (d), redox-age class was assigned to 5 percent of the 6,593 samples used in this study.

Overall, samples classified as oxic-potentially young showed the largest changes in nitrate concentrations for pairs of samples collected at near decadal time scales, whereas samples classified as anoxic-potentially old showed the smallest changes (fig. 2). This result is consistent with what would be predicted on the basis of the discussions in the "Redox Classification" and "Groundwater-Age Classification" sections of this report. Changes in nitrate concentrations for the samples classified as oxic-potentially old and anoxic-potentially young were intermediate in scale. Despite the relatively large uncertainty in age that could be associated with the oxic-potentially old classification based on age criterion (d), it actually had a smaller interquartile range for changes in nitrate concentrations than the same redox-age class based on age criterion (c) (fig. 2).

Although nitrate concentration was one of the criteria used to assign groundwater-age classes (table 1), it is unlikely that using nitrate-concentration data in this manner biased the comparison between redox-age classes and changes in nitrate concentrations in paired samples (fig. 2). Age classes were assigned to just one of the paired samples (typically the more recently collected sample), and then compared to *changes* in nitrate concentrations. A high nitrate concentration in a single groundwater sample may be an indicator of recently impacted groundwater, and that is the point of using it as an age indicator, but it is not a guarantee that changes in nitrate concentrations occurred at the time scale of the trends sampling.

In general, only the more recently collected sample was assigned a redox-age class for pairs of samples collected at near decadal time scales. An attempt was made to assign redox-age classes to both pairs of samples, but it quickly became apparent that the age assignments would not be comparable in most cases. This is because the paired samples usually were not analyzed for the same suite of pesticide compounds and VOCs, and often only one of the samples was analyzed for tritium.

Because data for dissolved oxygen typically were available for both pairs of samples, changes in nitrate concentrations were compared to changes in redox classification. The largest changes in nitrate concentrations occurred in pairs of samples that were both classified as oxic (fig. 3). This indicates that most of the changes in

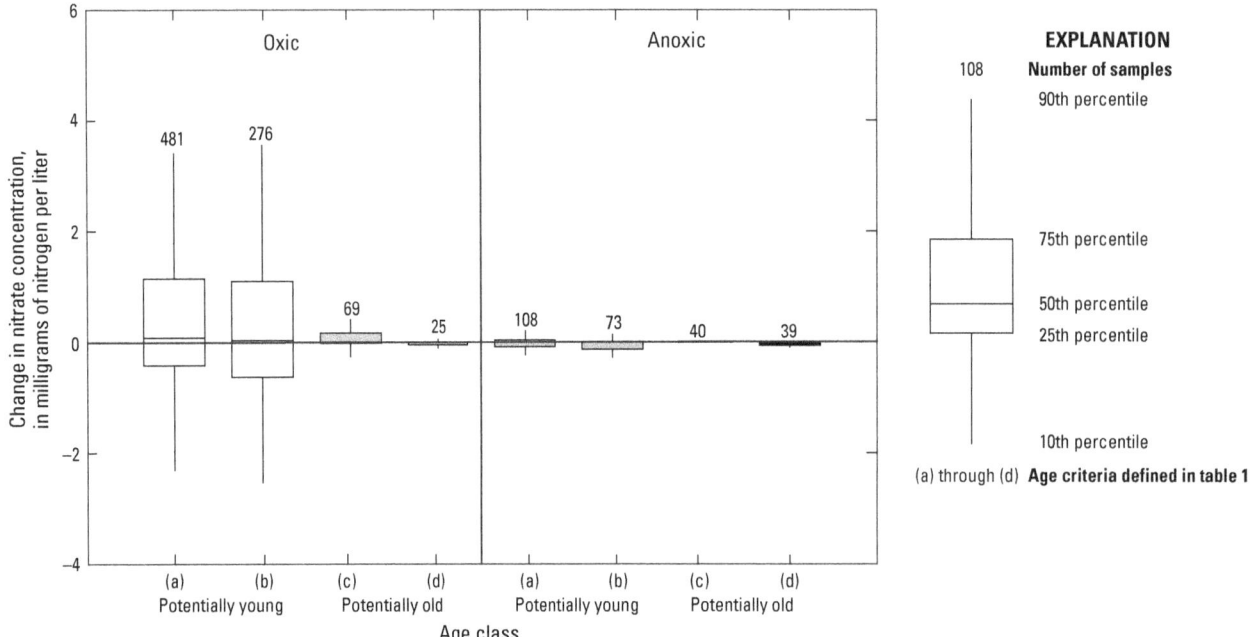

Figure 2. Change in nitrate concentration for pairs of samples collected from selected wells in the United States at near decadal time scales in relation to the redox-age class of the more recently collected sample.

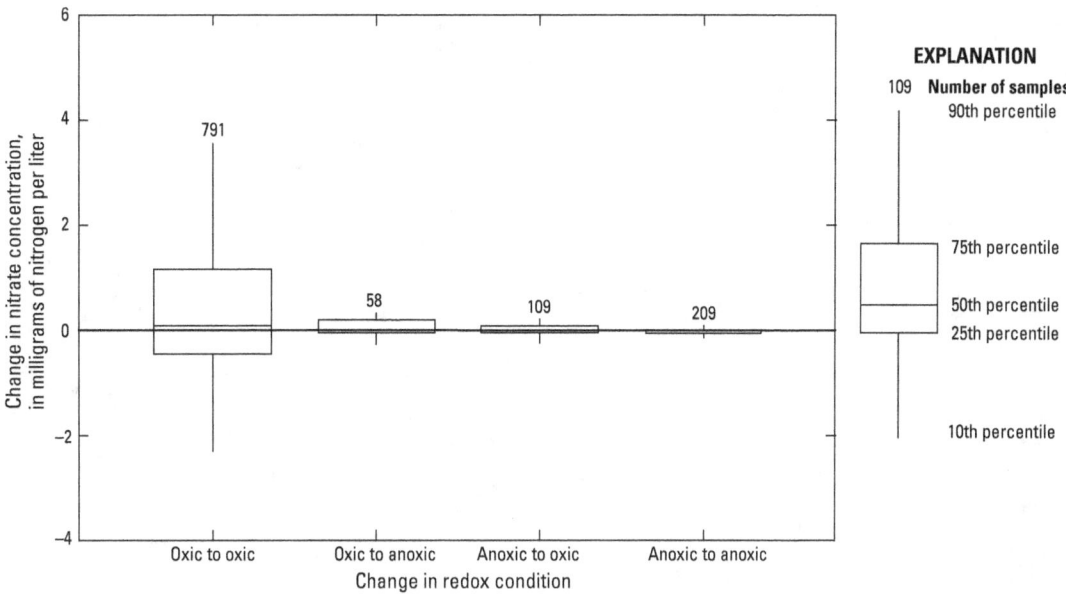

Figure 3.   Change in nitrate concentration for pairs of samples collected from selected wells in the United States at near decadal time scales in relation to the change in redox condition.

nitrate concentrations were not a result of changes in redox conditions in the aquifer but were more likely a result of changing nitrogen inputs at the land surface and (or) changing fractions of young and old water in the sample pairs.

On the basis of the evaluation presented above, the redox-age classes in table 1 can provide useful information on aquifer susceptibility to changes in nitrate concentrations. For the remainder of this report, results for age criteria (a) and (b) were combined by redox class and the same was done for age criteria (c) and (d). This results in four redox-age classes (oxic-potentially young, oxic-potentially old, anoxic-potentially young, and anoxic-potentially old) instead of eight.

Relation Between Redox-Age Classes and Changes in Nitrate Concentrations in Trend-Well Networks

The data in figure 2 indicate a strong relation between redox-age class and the change in nitrate concentration in pairs of samples collected from individual wells at near decadal time scales. The relation between redox-age classes and changes in nitrate concentrations also was examined at the well-network level because the well networks, unlike single wells, were designed to be statistically representative of large aquifer areas. Lindsey and Rupert (2012) analyzed nitrate concentrations in water samples collected at near

decadal time scales from 56 NAWQA well networks (fig. 4) and found statistically significant changes in concentrations at greater than a 90-percent confidence level in 18 (32 percent) of them. Lindsey and Rupert (2012) did not analyze nitrogen input histories in those 18 networks, but presumably the significant changes in nitrate concentrations were related to changes in nitrogen inputs at the land surface in some of them. Other factors such as variations in recharge rates, depth to groundwater, or pumping also could have affected nitrate concentrations (Rosen and others, 2008). Redox-age classes assigned to water samples from the LUS and MAS networks analyzed by Lindsey and Rupert (2012) are shown in table 3.

Results from 6 of 25 agricultural LUS networks for which near decadal changes could be evaluated showed significant increases in nitrate concentrations and 2 networks showed significant decreases (Lindsey and Rupert, 2012) (fig. 5 and table 3). Agricultural networks that showed significant increases in concentrations were located in the Central Valley aquifer system, Central glacial aquifers, Floridan aquifer system, Snake River Plain basaltic-rock aquifer system, and the South Platte River alluvial aquifer (fig. 4 and table 3). At least 75 percent of the samples in 7 of the 8 networks that showed significant changes in nitrate concentrations were classified as oxic-potentially young (fig. 5), and no more than 4 percent of the samples were classified as anoxic-potentially old (table 3). Samples from urban LUSs generally showed similar results, with 4 of 13 networks showing significant increases in nitrate concentrations and 1 showing a significant decrease. At least

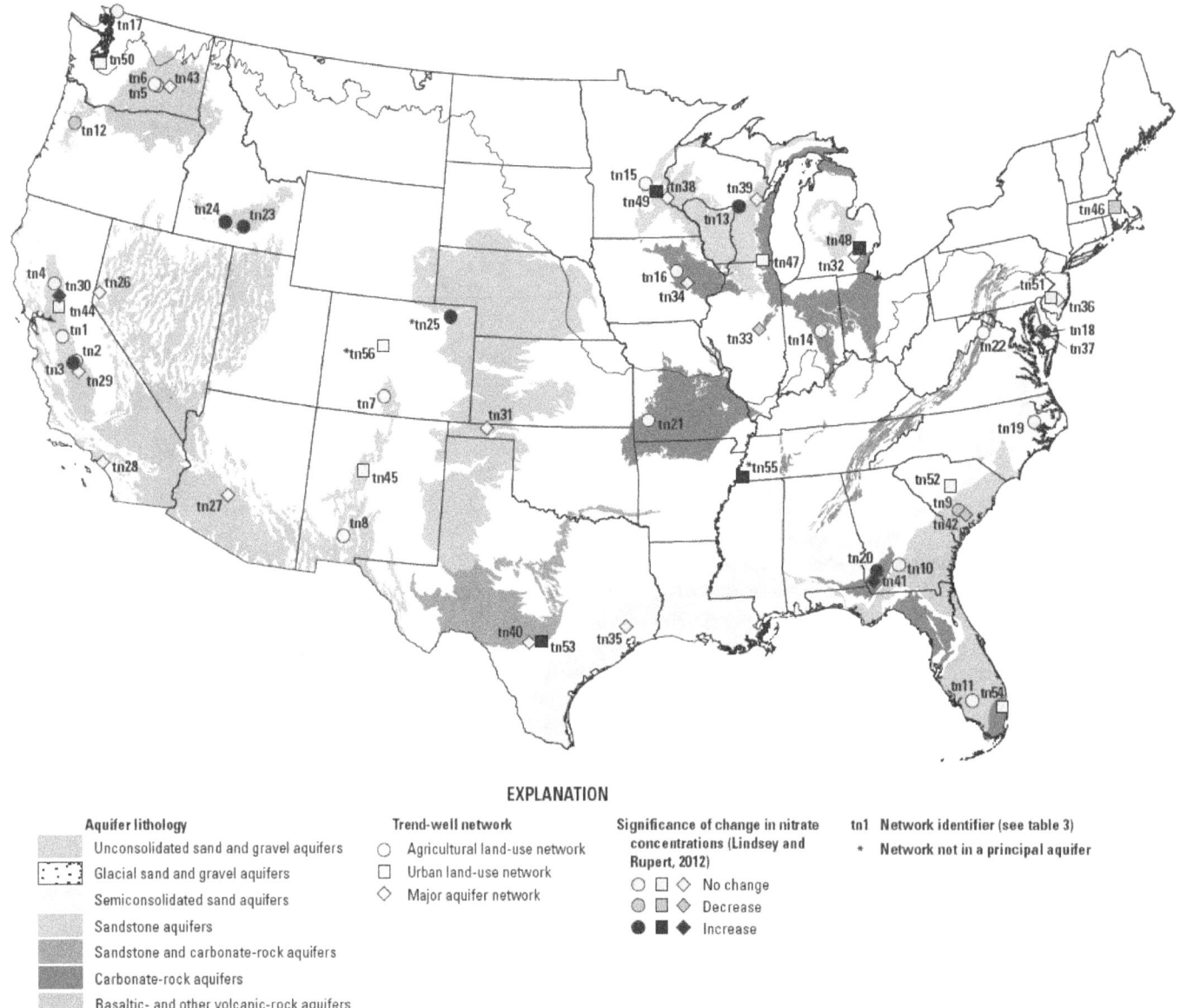

Aquifer lithology

Unconsolidated sand and gravel aquifers

Glacial sand and gravel aquifers

Semiconsolidated sand aquifers

Sandstone aquifers

Sandstone and carbonate-rock aquifers

Carbonate-rock aquifers

Basaltic- and other volcanic-rock aquifers

Trend-well network

○ Agricultural land-use network

□ Urban land-use network

◇ Major aquifer network

Significance of change in nitrate concentrations (Lindsey and Rupert, 2012)

○ □ ◇ No change

◔ ◱ ◈ Decrease

● ■ ◆ Increase

tn1   Network identifier (see table 3)

*   Network not in a principal aquifer

Figure 4.   Central locations of well networks sampled at near decadal time scales by the National Water-Quality Assessment Program.

75 percent of the samples in 4 of the 5 networks that showed significant changes in nitrate concentrations were classified as oxic-potentially young (fig. 5), and no more than 5 percent of the samples were classified as anoxic-potentially old (table 3). For 11 of the 17 agricultural LUSs and 7 of the 8 urban LUSs that showed no significant changes in nitrate concentrations, at least 75 percent of their samples also were classified as oxic-potentially young (table 3). Only about 11 percent of the LUS networks studied by Lindsey and Rupert (2012) had more than 10 percent of their samples classified as potentially old (table 3), which is not surprising given that NAWQA land-use studies typically targeted the most recently recharged groundwater.

Three of 18 MASs showed significant increases in nitrate concentrations, and two showed significant decreases (Lindsey and Rupert, 2012) (fig. 4 and table 3). Networks that showed

significant increases in nitrate concentrations were located in the Central Valley, Northern Atlantic Coastal Plain, and Floridan aquifer systems. Only one of the MASs (acfbsus1) that showed a significant change in nitrate concentrations had more than 75 percent of the samples classified as oxic-potentially young. For the other four networks, 0 to 54 percent of the samples were classified as oxic-potentially young (fig. 5 and table 3). On closer inspection, the LUS and MAS results are not necessarily inconsistent because most of the wells with large changes in nitrate concentrations in these MAS networks with significant changes were oxic-potentially young. For the two MAS networks (lirbsus1 and santsus2) that showed significant decreases in nitrate concentrations, 62 to 100 percent of the samples were classified as anoxic and 69 to 79 percent of the samples were classified as potentially old (table 3). The network-level changes in nitrate concentrations for lirbsus1 and santsus2 were –0.04 and –0.05 mg-N/L, respectively (fig. 5 and table 3), and the change in concentration for almost all the sample pairs was less than 0.1 mg-N/L (figs. 6A, B). More than 50 percent of the data were pairs of nondetects (Lindsey and Rupert, 2012). The one sample in those two networks than did show a relatively large change in concentration (greater than 1 mg-N/L) was classified as oxic-potentially young (fig. 6B). The predominance of small changes in nitrate concentrations in these two networks would be expected for aquifers that contained large percentages of anoxic and (or) potentially old water. The two networks (sacrsus1 and dlmvsus1) that showed significant increases in nitrate concentrations, but for which the percentage of samples classified as oxic-potentially young was less than 75 percent (48 to 54 percent), had larger absolute changes in nitrate concentrations (0.14 mg-N/L) than lirbsus1 and santsus2 (fig. 5). The concentration changes for sacrsus1 and dlmvsus1, however, were smaller than the change for the MAS network acfbsus1 (0.32 mg-N/L) that had more than 75 percent of its samples classified as oxic-potentially young (fig. 5 and table 3). Closer inspection of the data from sacrsus1 and dlmvsus1 shows that most of the large changes in concentrations occurred in samples that were classified as oxic-potentially young (figs. 6C, D). Three samples from the dlmvsus1 study that were classified as anoxic-potentially young showed concentration changes of 4.9 to 7.5 mg-N/L (fig. 6D); redox conditions in these samples apparently were anoxic but did not result in complete denitrification.

Two of the 13 MASs that showed no significant changes in nitrate concentrations had at least 75 percent of their samples classified as oxic-potentially young (table 3). The remaining 11 networks had 32 to 71 percent of their samples classified as oxic-potentially young. About 83 percent of the MAS networks studied by Lindsey and Rupert (2012) had more than 10 percent of their samples classified as potentially old, which is a considerably larger percentage than for the agricultural LUS networks. This finding was expected because wells used in the MAS networks typically were deeper than those used in the LUS networks.

The redox-age results for pairs of samples collected from individual wells show that the largest changes in nitrate concentrations primarily occurred in samples that were classified as oxic-potentially young (figs. 2 and 6). A generally similar pattern was observed when samples were aggregated to the level of well networks. For LUS and MAS networks that showed significant changes in nitrate concentrations, the median changes in concentrations were 0.28, 0.14, and –0.05 mg-N/L for networks that had at least 75 percent, at least 25 percent and less than 75 percent, and less than 25 percent of the samples classified as oxic-potentially young, respectively. On the basis of the data shown in figures 2 and 5, aquifer areas (as defined by well networks) in which at least 75 percent of the samples were classified as oxic-potentially young were considered to have a high susceptibility to

changes in nitrate concentrations (fig. 5). Aquifer areas were considered to have a medium susceptibility to changes in nitrate concentrations if at least 25 percent and less than 75 percent of the samples were classified as oxic-potentially young (fig. 5). Aquifer areas were considered to have a low susceptibility to changes in nitrate concentrations if less than 25 percent of the samples were classified as oxic-potentially young (fig. 5). These definitions of high, medium, and low are used to characterize aquifer susceptibility to changes in nitrate concentrations throughout the remainder of the report. The degree of susceptibility is not intended to indicate that significant changes in concentrations of nitrate will or will not be detected in the future in those areas. Other factors such as nitrogen input history at the land surface, mixing, and lag times related to nitrate transport could cause nitrate-concentration trends to develop over longer time scales than the near decadal time scale examined by Lindsey and Rupert (2012), or not at all.

Redox-Age Classes in Principal Aquifers

The three primary well types sampled by NAWQA generally represent different depth zones and (or) areas of the principal aquifers. Shallow monitoring wells are completed near the water table, whereas domestic and public-supply wells are mostly completed in deeper zones in the aquifers. In this section of the report, redox-age classes assigned to networks of shallow monitoring wells in agricultural areas, domestic wells, and public-supply wells were used to characterize the susceptibility to changes in nitrate concentrations of these different depth zones and (or) areas of the principal aquifers. The susceptibility to changes in nitrate concentrations near the water table in urban areas is not considered here because of the generally small area represented by those networks of shallow monitoring wells, however, redox-age classes and susceptibility rankings for those networks are shown in the Appendix.

Susceptibility to Changes in Nitrate Concentrations Near the Water Table in Agricultural Areas

Redox-age classes were assigned to samples collected from 40 networks of shallow monitoring wells in agricultural areas (fig. 7 and table 4). Most networks (58 percent) had a high susceptibility to changes in nitrate concentrations because at least 75 percent of their samples were classified as oxic-potentially young. Only 10 percent of the networks had a low susceptibility to changes in nitrate concentrations because less than 25 percent of their samples classified as oxic-potentially young (table 4).

Table 3.   Redox-age classes for water samples collected from selected well networks in the United States at near decadal time scales, median change in nitrate concentration for each network, and the statistical significance of the change in concentration.

[usg, unconsolidated sand and gravel; gla, glacial sand and gravel; scs, semiconsolidated sand; car, carbonate rock; bav, basaltic and other volcanic rock; san, sandstone; scr, sandstone and carbonate rock; alus, agricultural land-use study; mas, major-aquifer study; ulus, urban land-use study; mg-N/L, milligrams of nitrogen per liter; shading is used to differentiate between alus, mas, and ulus studies; bold indicates a statistically significant change in nitrate concentrations at greater than a 90-percent confidence level]

Aquifer number	Aquifer lithology	Aquifer name	Study type	Network name (number of wells)	Network identifier (see figure 4)
3	usg	Central Valley aquifer system	alus	sanjlusor2a (19)	tn1
3	usg	Central Valley aquifer system	alus	sanjlusor1a (17)	tn2
3	**usg**	**Central Valley aquifer system**	**alus**	**sanjluscr1a (18)**	**tn3**
3	usg	Central Valley aquifer system	alus	sacrluscr1 (21)	tn4
4	usg	Columbia Plateau basin-fill aquifers	alus	ccptlusag2b (16)	tn5
4	usg	Columbia Plateau basin-fill aquifers	alus	ccptlusor1b (19)	tn6
8	usg	Rio Grande aquifer system	alus	riogluscr1 (12)	tn7
8	usg	Rio Grande aquifer system	alus	rioglusag1 (25)	tn8
10	**usg**	**Surficial aquifer system**	**alus**	**santluscr1 (19)**	**tn9**
10	usg	Surficial aquifer system	alus	gaflluscr1 (20)	tn10
10	usg	Surficial aquifer system	alus	sofllusor1 (17)	tn11
11	**usg**	**Willamette Lowland aquifer system**	**alus**	**willlusag3 (24)**	**tn12**
12c	**gla**	**Central glacial aquifers**	**alus**	**wmiclusag2 (26)**	**tn13**
12c	gla	Central glacial aquifers	alus	whitluscr1 (20)	tn14
12wc	gla	West-central glacial aquifers	alus	umisluscr1 (22)	tn15
12wc	gla	West-central glacial aquifers	alus	eiwaluscr1 (30)	tn16
12w	gla	Western glacial aquifers	alus	pugtluscr1 (19)	tn17
15	scs	Northern Atlantic Coastal Plain aquifer system	alus	dlmvluscr1 (16)	tn18
15	scs	Northern Atlantic Coastal Plain aquifer system	alus	albelusag1 (12)	tn19
29	**car**	**Floridan aquifer system**	**alus**	**acfbluscr3 (19)**[1]	**tn20**
31	car	Ozark Plateaus aquifer system	alus	ozrklusag2a (20)	tn21
34	car	Valley and Ridge carbonate-rock aquifers	alus	potolusag1 (24)	tn22
36	**bav**	**Snake River Plain basaltic-rock aquifer system**	**alus**	**usnkluscr2 (26)**	**tn23**
36	**bav**	**Snake River Plain basaltic-rock aquifer system**	**alus**	**usnkluscr3 (28)**	**tn24**
--[3]	**usg**	**South Platte River alluvial aquifer**	**alus**	**spltluscr1 (29)**	**tn25**
1	usg	Basin and Range basin-fill aquifers	mas	nvbrsus2 (16)	tn26
1	usg	Basin and Range basin-fill aquifers	mas	cazbsus1a (24)	tn27
2	usg	California Coastal Basin aquifers	mas	sanasus2 (14)	tn28
3	usg	Central Valley aquifer system	mas	sanjsus1 (26)	tn29
3	**usg**	**Central Valley aquifer system**	**mas**	**sacrsus1 (28)**	**tn30**
5	usg	High Plains aquifer	mas	hpgwsus1a (30)	tn31
12c	gla	Central glacial aquifers	mas	lerisus1 (27)	tn32
12c	**gla**	**Central glacial aquifers**	**mas**	**lirbsus1 (29)**	**tn33**
12wc	gla	West-central glacial aquifers	mas	eiwasus2 (30)	tn34
13	scs	Coastal Lowlands aquifer system	mas	trinsus1 (17)	tn35
15	scs	Northern Atlantic Coastal Plain aquifer system	mas	linjsus2 (24)	tn36
15	**scs**	**Northern Atlantic Coastal Plain aquifer system**	**mas**	**dlmvsus1 (23)**	**tn37**
18	san	Cambrian-Ordovician aquifer system	mas	umissus3 (22)	tn38
18	san	Cambrian-Ordovician aquifer system	mas	wmicsus1 (25)	tn39
24	scr	Edwards-Trinity aquifer system	mas	sctxsus1 (23)	tn40
29	**car**	**Floridan aquifer system**	**mas**	**acfbsus1 (20)**	**tn41**
29	**car**	**Floridan aquifer system**	**mas**	**santsus2 (29)**	**tn42**
35	bav	Columbia Plateau basaltic-rock aquifer system	mas	ccptsus1b (30)[2]	tn43
3	usg	Central Valley aquifer system	ulus	sacrlusrc1 (18)	tn44
8	usg	Rio Grande aquifer system	ulus	rioglusrc1 (10)	tn45
12e	**gla**	**Eastern glacial aquifers**	**ulus**	**necblusrc1 (21)**	**tn46**
12c	gla	Central glacial aquifers	ulus	uirblusrc1 (18)	tn47
12c	**gla**	**Central glacial aquifers**	**ulus**	**lerilusrc1 (29)**	**tn48**
12wc	**gla**	**West-central glacial aquifers**	**ulus**	**umislusrc1 (26)**	**tn49**
12w	gla	Western glacial aquifers	ulus	pugtlusrs1 (24)	tn50
15	scs	Northern Atlantic Coastal Plain aquifer system	ulus	linjlusrc1 (27)	tn51
16	scs	Southeastern Coastal Plain aquifer system	ulus	santlusrc1 (17)	tn52
24	**scr**	**Edwards-Trinity aquifer system**	**ulus**	**sctxlusrc1 (30)**	**tn53**
27	car	Biscayne aquifer	ulus	sofllusrc1a (17)	tn54
--[3]	**usg**	**Alluvial aquifer in Memphis, Tennessee**	**ulus**	**miselusrc1 (20)**	**tn55**
--[3]	usg	Alluvial aquifers in the Colorado Rocky Mountains	ulus	ucollusrc1 (16)	tn56

Table 3. Redox-age classes for water samples collected from selected well networks in the United States at near decadal time scales, median change in nitrate concentration for each network, and the statistical significance of the change in concentration.—Continued

[usg, unconsolidated sand and gravel; gla, glacial sand and gravel; scs, semiconsolidated sand; car, carbonate rock; bav, basaltic and other volcanic rock; san, sandstone; scr, sandstone and carbonate rock; alus, agricultural land-use study; mas, major-aquifer study; ulus, urban land-use study; mg-N/L, milligrams of nitrogen per liter; shading is used to differentiate between alus, mas, and ulus studies; bold indicates a statistically significant change in nitrate concentrations at greater than a 90-percent confidence level]

Aquifer number	Redox-age class (percentage of samples)[4]				Data from Lindsey and Rupert (2012)	
	Oxic-potentially young	Oxic-potentially old	Anoxic-potentially young	Anoxic-potentially old	Median change in nitrate concentration (mg-N/L)	Statistical significance of change in nitrate concentration
3	95	0	5	0	0.62	No change
3	88	6	6	0	0.07	No change
3	**82**	**0**	**18**	**0**	**1.0**	**Increase**
3	10	0	86	5	−0.04	No change
4	94	6	0	0	−0.77	No change
4	89	0	11	0	0.05	No change
8	75	8	17	0	−0.11	No change
8	40	8	48	4	0.13	No change
10	**95**	**0**	**5**	**0**	**−0.63**	**Decrease**
10	90	10	0	0	0.84	No change
10	41	0	53	6	−0.02	No change
11	**46**	**0**	**54**	**0**	**−0.04**	**Decrease**
12c	**85**	**0**	**12**	**4**	**0.84**	**Increase**
12c	45	0	50	5	0.00	No change
12wc	100	0	0	0	0.88	No change
12wc	97	3	0	0	−0.02	No change
12w	79	0	21	0	1.1	No change
15	56	0	44	0	−0.10	No change
15	33	0	58	8	0.03	No change
29	**89**	**11**	**0**	**0**	**0.29**	**Increase**
31	81	0	19	0	0.02	No change
34	96	0	4	0	−0.09	No change
36	**100**	**0**	**0**	**0**	**0.26**	**Increase**
36	**96**	**4**	**0**	**0**	**0.06**	**Increase**
--[3]	**83**	**0**	**17**	**0**	**2.0**	**Increase**
1	69	19	0	13	0.05	No change
1	38	63	0	0	0.14	No change
2	50	7	21	21	−0.05	No change
3	81	8	4	8	0.45	No change
3	**54**	**18**	**25**	**4**	**0.14**	**Increase**
5	43	57	0	0	0.01	No change
12c	33	22	19	26	0.01	No change
12c	**0**	**0**	**21**	**79**	**−0.04**	**Decrease**
12wc	67	27	3	3	−0.04	No change
13	35	59	6	0	0.00	No change
15	71	0	13	17	−0.01	No change
15	**48**	**0**	**52**	**0**	**0.14**	**Increase**
18	55	9	32	5	−0.04	No change
18	32	4	36	28	0.01	No change
24	91	0	9	0	−0.04	No change
29	**100**	**0**	**0**	**0**	**0.32**	**Increase**
29	**10**	**28**	**21**	**41**	**−0.05**	**Decrease**
35	69	7	3	21	0.02	No change
3	83	0	17	0	−0.03	No change
8	30	0	30	40	0.05	No change
12e	**90**	**0**	**10**	**0**	**−0.09**	**Decrease**
12c	93	0	7	0	−0.01	No change
12c	**72**	**0**	**28**	**0**	**0.28**	**Increase**
12wc	**88**	**0**	**12**	**0**	**0.04**	**Increase**
12w	92	0	4	4	0.18	No change
15	93	0	7	0	−0.09	No change
16	82	12	6	0	0.04	No change
24	**100**	**0**	**0**	**0**	**0.36**	**Increase**
27	82	6	12	0	0.02	No change
--[3]	**75**	**0**	**20**	**5**	**0.21**	**Increase**
--[3]	75	6	19	0	−0.03	No change

[1]Network has wells in the Floridan and Southeastern Coastal Plain aquifer systems.

[2]Network has wells in the Columbia Plateau basaltic-rock and basin-fill aquifer systems.

[3]Network not in a principal aquifer.

[4]Redox-age classes where determined for the more recently collected samples in each network, and redox-age percentages may not sum to 100 percent because of rounding.

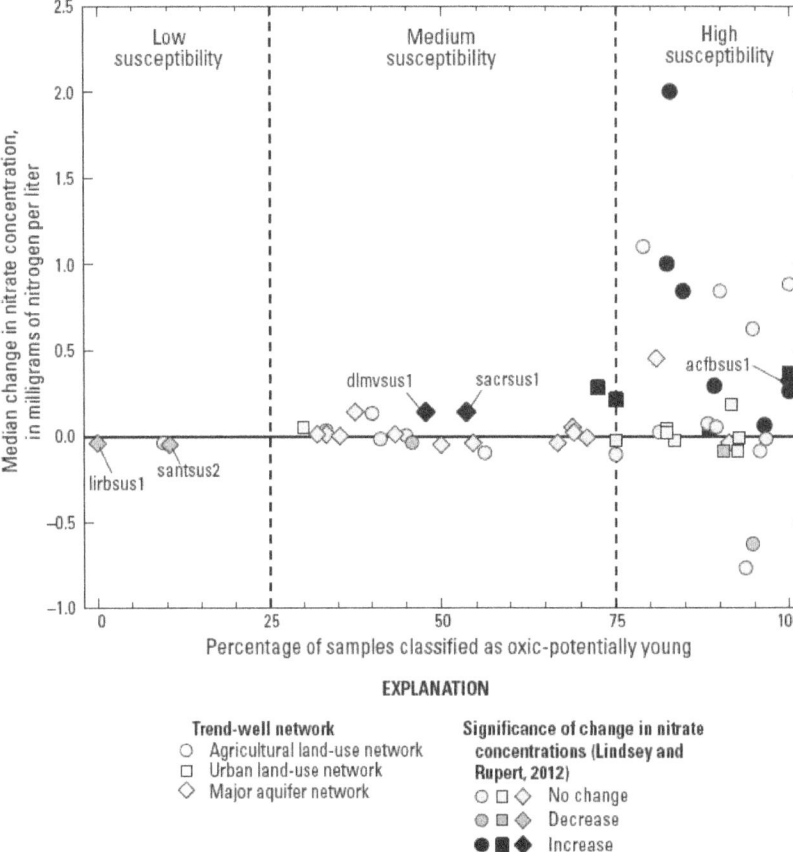

Figure 5.   Median change in nitrate concentration in relation to the percentage of samples that were classified as oxic-potentially young in well networks that were sampled at near decadal timescales (concentration data from Lindsey and Rupert, 2012), and the susceptibility of the networks to changes in nitrate concentrations.

For principal aquifers that had at least 2 networks of wells, median percentages of samples classified as oxic-potentially young ranged from about 57 to 96 percent (fig. 8). On the basis of these data, for the parts of aquifers near the water table in agricultural areas, the aquifers most susceptible to changes in nitrate concentrations were the Columbia Plateau basin-fill aquifers, Eastern glacial aquifers, and the West-central glacial aquifers (fig. 8). None of the aquifers had a low susceptibility to changes in nitrate concentrations, which would be indicated by a median percentage of samples classified as oxic-potentially young that was less than 25 percent. The High Plains aquifer had the highest median percentage of samples classified as oxic-potentially old, which generally reflects the low organic-carbon content of sediment and relatively low recharge rates in the aquifer (McMahon and others, 2007). The Central Valley and Surficial aquifer systems had the highest median percentages of samples classified as anoxic-potentially young (fig. 8). Only the Rio Grande aquifer system and the Central glacial aquifers had median percentages of samples classified as anoxic-potentially old that were greater than zero.

Although the median percentages in figure 8 provide a general comparison of redox-age classes between principal aquifers, they do not indicate the substantial redox-age variability that can occur within an aquifer. For the three well networks in the Central Valley aquifer system, the percentage of samples classified as oxic-potentially young ranged from 13 to 90 percent (table 4 and fig. 9), and the percentage of samples classified as anoxic-potentially young ranged from 10 to 83 percent. In the High Plains aquifer, median percentages of samples classified as oxic-potentially old ranged from 10 to 41 percent (fig. 9). In the West-central glacial aquifer, median percentages of samples classified as anoxic-potentially old ranged from 0 to 35 percent (fig. 9). Large intraaquifer redox-age variability was observed in most of the aquifers that had multiple networks of shallow monitoring wells in agricultural areas (table 4).

Distinct patterns in the spatial distribution of network-level susceptibilities are apparent in some of the aquifers. Networks in the Central glacial aquifers of Indiana and parts of southern Michigan and Wisconsin had medium susceptibilities, whereas networks in glacial aquifers to the

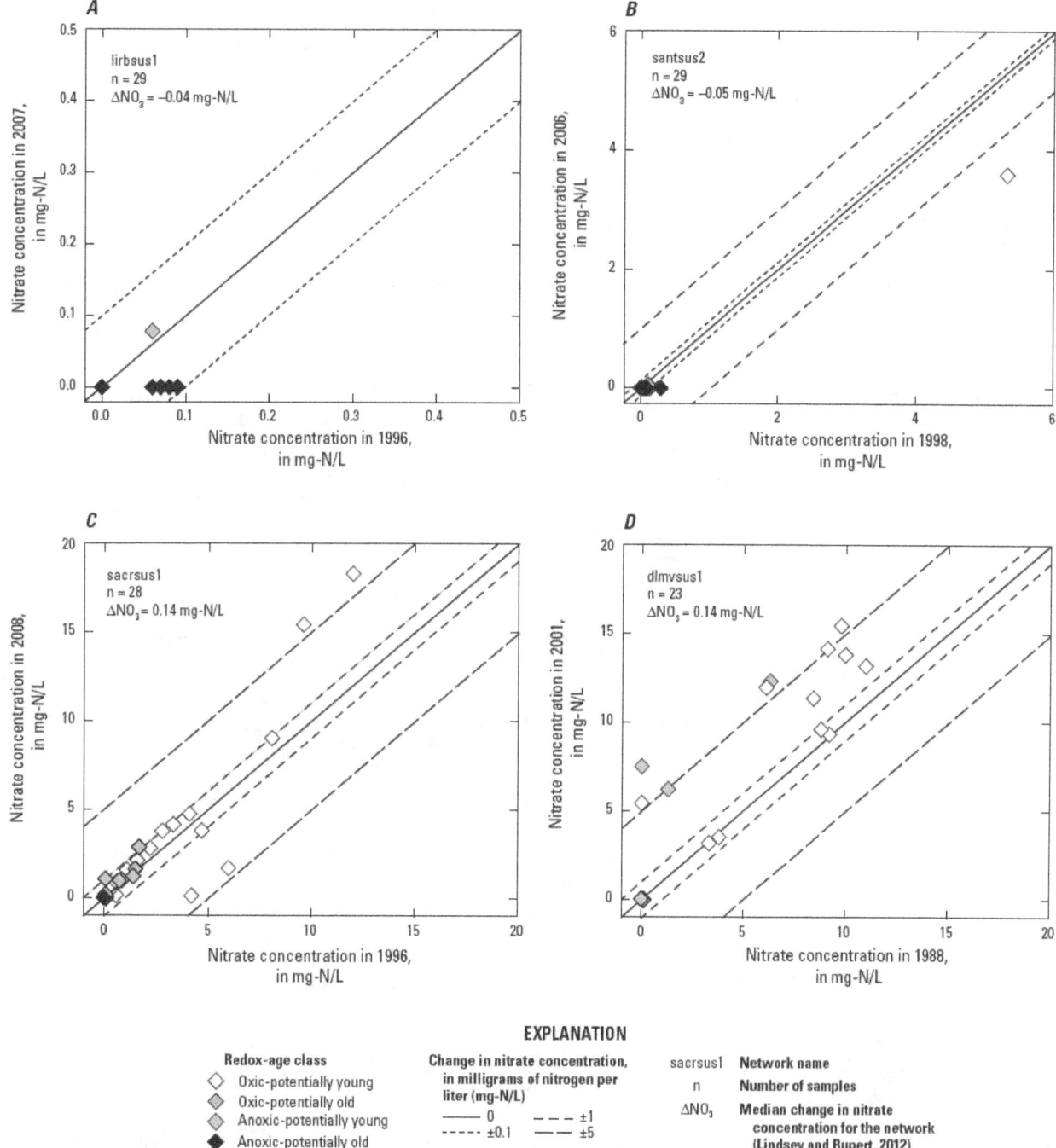

EXPLANATION

Redox-age class
◇ Oxic-potentially young
◆ Oxic-potentially old
◆ Anoxic-potentially young
◆ Anoxic-potentially old

Change in nitrate concentration, in milligrams of nitrogen per liter (mg-N/L)
—— 0
----- ±0.1
— — ±1
— — ±5

sacrsus1   Network name
n   Number of samples
ΔNO₃   Median change in nitrate concentration for the network (Lindsey and Rupert, 2012)

Figure 6.   Concentrations of nitrate in pairs of samples collected from selected networks of major-aquifer study wells in the United States at near decadal time scales; (*A*) lirbsus1, (*B*) santsus2, (*C*) sacrsus1, and (*D*) dlmvsus1 networks.

Figure 7. Central locations of networks of shallow monitoring wells in agricultural areas and the susceptibility of the networks to changes in nitrate concentrations.

Table 4. Redox-age classes for water samples collected from networks of shallow monitoring wells in agricultural areas in the United States and the susceptibility of the networks to changes in nitrate concentrations (only networks with at least 10 wells are listed).

[usg, unconsolidated sand and gravel; gla, glacial sand and gravel; scs, semiconsolidated sand; scr, sandstone and carbonate rock; car, carbonate rock; shading is used to differentiate between aquifer lithologies]

Aquifer number	Aquifer lithology	Aquifer name	Network name (number of wells)	Network identifier (see figure 7)	Redox-age class (percentage of samples)[3]				Susceptibility to changes in nitrate concentrations
					Oxic-potentially young	Oxic-potentially old	Anoxic-potentially young	Anoxic-potentially old	
1	usg	Basin and Range basin-fill aquifers	nvbrlusag1 (13)	am1	15	15	31	38	Low
3	usg	Central Valley aquifer system	sanjlusor1b (10)	am2	90	0	10	0	High
3	usg	Central Valley aquifer system	sanjluscr1b (10)	am3	60	0	40	0	Medium
3	usg	Central Valley aquifer system	sacrluscr1 (30)	am4	13	0	83	3	Low
4	usg	Columbia Plateau basin-fill aquifers	ccptlusag2b (27)	am5	93	4	4	0	High
4	usg	Columbia Plateau basin-fill aquifers	ccptlusor1b (25)	am6	88	0	12	0	High
5	usg	High Plains aquifer	hpgwlusag3 (30)	am7	90	10	0	0	High
5	usg	High Plains aquifer	hpgwlusag1 (27)	am8	74	26	0	0	Medium
5	usg	High Plains aquifer	hpgwlusag2 (27)	am9	59	41	0	0	Medium
8	usg	Rio Grande aquifer system	riogluscr1 (76)	am10	87	1	12	0	High
8	usg	Rio Grande aquifer system	rioglusag1 (34)	am11	32	9	50	9	Medium
10	usg	Surficial aquifer system	santluscr1 (30)	am12	93	0	7	0	High
10	usg	Surficial aquifer system	sofllusor1 (38)	am13	21	0	76	3	Low
12e	gla	Eastern glacial aquifers	hdsnlusag1 (12)	am14	100	0	0	0	High
12e	gla	Eastern glacial aquifers	comnlusag1 (32)	am15	91	3	6	0	High
12c	gla	Central glacial aquifers	lirbluscr1 (22)	am16	95	0	5	0	High
12c	gla	Central glacial aquifers	lirbluscr2 (25)	am17	88	0	12	0	High
12c	gla	Central glacial aquifers	wmiclusag2 (29)	am18	86	0	10	3	High
12c	gla	Central glacial aquifers	miamluscr1 (21)	am19	86	0	14	0	High
12c	gla	Central glacial aquifers	wmiclusag1a (23)	am20	74	4	13	9	Medium
12c	gla	Central glacial aquifers	whitluscr2 (20)	am21	65	10	20	5	Medium
12c	gla	Central glacial aquifers	uirbluscr1 (29)	am22	59	0	41	0	Medium
12c	gla	Central glacial aquifers	leriluscr1 (30)	am23	57	13	30	0	Medium
12c	gla	Central glacial aquifers	whitluscr3a (24)	am24	50	0	46	4	Medium
12c	gla	Central glacial aquifers	whitluscr1 (26)	am25	46	0	50	4	Medium
12wc	gla	West-central glacial aquifers	umisluscr1 (27)	am26	100	0	0	0	High
12wc	gla	West-central glacial aquifers	eiwaluscr1 (31)	am27	97	3	0	0	High
12wc	gla	West-central glacial aquifers	rednlusag1 (21)	am28	90	0	10	0	High
12wc	gla	West-central glacial aquifers	cnbrluscr1 (26)	am29	88	0	12	0	High
12wc	gla	West-central glacial aquifers	rednlusag2 (20)	am30	15	5	45	35	Low
12w	gla	Western glacial aquifers	pugtluscr1 (20)	am31	75	0	25	0	High
13	scs	Coastal Lowlands aquifer system	acadluscr1 (21)	am32	100	0	0	0	High
15	scs	Northern Atlantic Coastal Plain aquifer system	linjluscr1 (15)	am33	100	0	0	0	High
15	scs	Northern Atlantic Coastal Plain aquifer system	dlmvluscr1 (27)	am34	67	0	33	0	Medium
15	scs	Northern Atlantic Coastal Plain aquifer system	albelusag1 (30)	am35	50	0	43	7	Medium
25	scr	Mississippian aquifers	ltenlusag1 (29)	am36	97	0	3	0	High
29	car	Floridan aquifer system	acflusscr3 (24)[1]	am37	92	8	0	0	High
--[2]	usg	Denver Basin alluvial aquifers	spltluscr2 (21)	am38	100	0	0	0	High
--[2]	usg	South Platte River alluvial aquifer	spltluscr1 (30)	am39	83	0	17	0	High
--[2]	usg	Valley and Ridge alluvial aquifers	utenluscr1 (30)	am40	63	0	33	3	Medium

[1]Network has wells in the Floridan and Southeastern Coastal Plain aquifer systems.

[2]Network not in a principal aquifer.

[3]Redox-age percentages may not sum to 100 percent because of rounding.

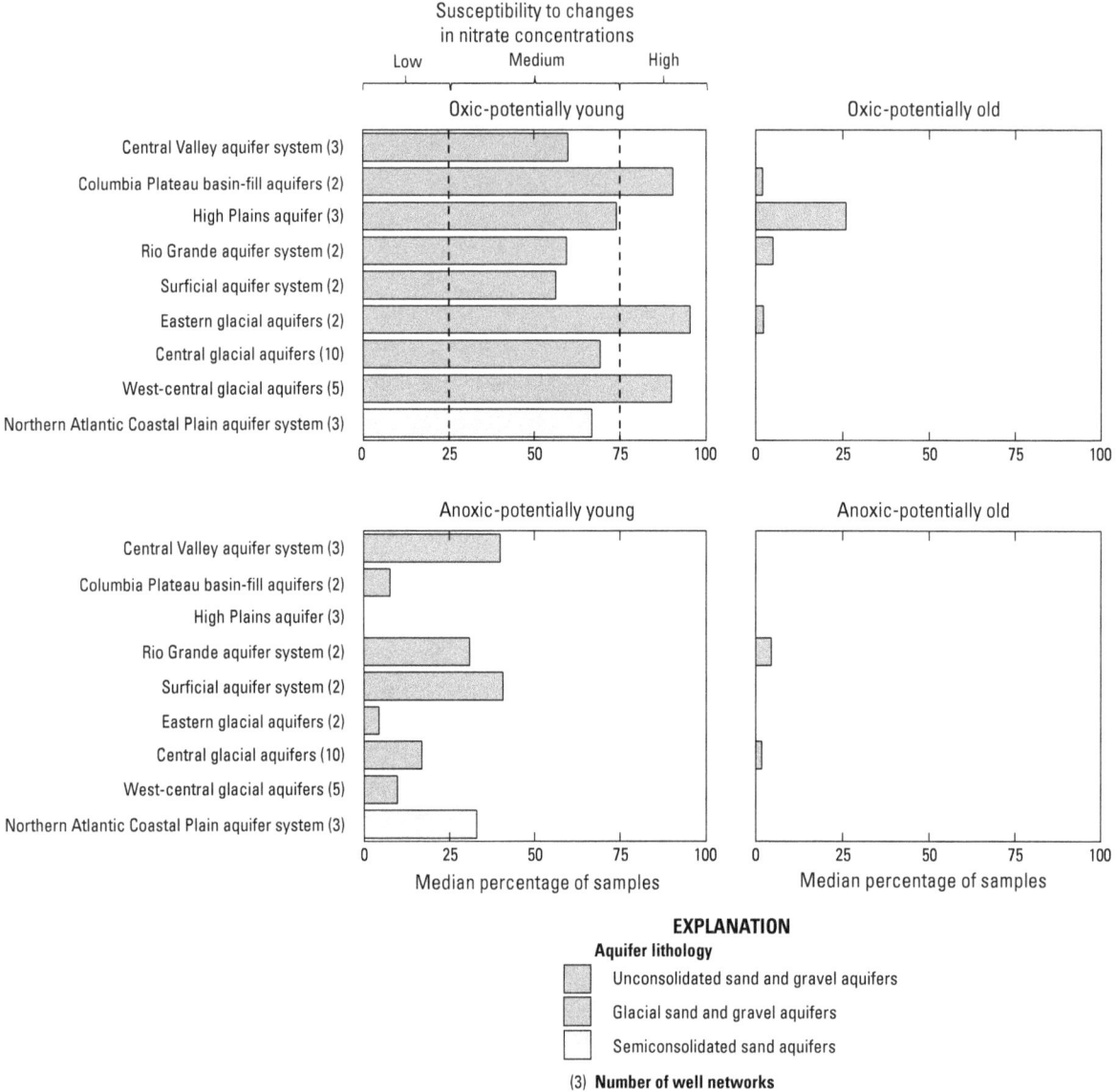

Figure 8.   Median percentage of samples assigned to the four redox-age classes for principal aquifers that have at least two networks of shallow monitoring wells in agricultural areas, and the susceptibility of the aquifers to changes in nitrate concentrations.

east and west had high susceptibilities. The areas of medium susceptibility in Indiana had relatively high percentages of samples classified as anoxic-potentially young compared to the surrounding networks with high susceptibilities (fig. 7 and table 4). This difference in redox-age classes between the two areas could indicate shallower depths to water and (or) finer grained sediment in the medium-susceptibility area relative to the high-susceptibility areas, both of which could result in anoxic groundwater. Other studies have reported an increase in concentrations of dissolved organic carbon, probably from the soil zone, and a decrease in concentrations of dissolved oxygen in groundwater with decreasing depths to the water table (Pabich and others, 2001; McMahon and Chapelle, 2008). In the High Plains aquifer, network susceptibility decreased from high in the north to medium in the central and southern parts of the aquifer (fig. 7). This change in

susceptibility corresponds to a north-to-south increase in the percentage of samples classified as oxic-potentially old (fig. 9 and table 4), an increase that is probably related to the north-to-south decrease in recharge and increase in depth to the water table (McMahon and others, 2007). In the Central Valley aquifer system, network susceptibility increased from low in the north to medium and high in the south (fig. 7). The percentage of samples classified as anoxic-potentially young in the northern network was about 2 to 8 times greater than the percentages in the southern networks where oxic-potentially young groundwater predominated (fig. 9 and table 4). The common occurrence of anoxic groundwater in the north may be related to the much shallower depths to the water table in the northern network of wells (median depth 1.1 m) than in the southern networks (median depths 14 to 20 m).

Susceptibility to Changes in Nitrate Concentrations in Parts of Aquifers that Provide Domestic Water Supplies

Redox-age classes were assigned to samples collected from 105 networks of domestic wells (fig. 10 and table 5). Thirty-one percent of the networks were considered to have a high susceptibility to changes in nitrate concentrations and 17 percent of the networks were considered to have low susceptibilities (table 5). In comparison, 58 percent of the networks of shallow monitoring wells in agricultural areas were considered to have a high susceptibility to changes in nitrate concentrations and 10 percent were considered to have a low susceptibility.

For principal aquifers that had at least 2 networks of domestic wells, the median percentage of samples classified as oxic-potentially young ranged from about 6 to 100 percent (fig. 11), compared to about 57 to 96 percent for the shallow monitoring wells (fig. 8). For the parts of aquifers that provide domestic water supplies, the aquifers most susceptible to changes in nitrate concentrations were the Northern Atlantic Coastal Plain aquifer system and the Early Mesozoic Basin, Valley and Ridge carbonate-rock, and Piedmont and Blue Ridge crystalline-rock aquifers in the eastern United States; the Ozark Plateaus aquifer system in parts of Missouri and Arkansas; and the Central Valley, Columbia Plateau basaltic-rock, and Snake River Plain basaltic-rock aquifer systems in the West (figs. 10 and 11). For this analysis, western states are considered to be those located west of Minnesota, Iowa, Missouri, Arkansas, and Louisiana. The least susceptible aquifers were the Texas Coastal Uplands and Denver Basin aquifer systems (figs. 10 and 11).

Relatively large intraaquifer variability in redox-age classes was observed in some of the principal aquifers. For the five well networks sampled in the Floridan aquifer system, the percentage of samples classified as oxic-potentially young ranged from 10 to 100 percent (fig. 12). Aquifer confinement probably is an important control on redox-age variability in the Floridan aquifer system. More than 90 percent of the wells in network santsus2 were completed in the confined part of the aquifer and only 10 percent of its samples were classified as oxic-potentially young (table 5). Only 20 percent of the wells in network acfbsus1 were completed in the confined part of the aquifer and 100 percent of its samples were classified as oxic-potentially young. The Central glacial aquifers also showed large redox-age variability (fig. 12), which could be attributed to the diversity of depositional environments represented by well networks in those aquifers. Wells in network uirbsus1 were completed in glacial-moraine sands and gravels and 67 percent of their samples were classified as oxic-potentially young. Wells in network uirbsus2 were completed in glacial-till deposits and 48 percent of their samples were classified as oxic-potentially young. Wells in network lirbsus1 were completed in confined buried-bedrock-valley deposits and 0 percent of their samples were classified as oxic-potentially young. Not all of the aquifers exhibited large variability in redox-age classes. Networks in the Columbia Plateau and Snake River Plain basaltic-rock aquifers had consistently high percentages of samples classified as oxic-potentially young (fig. 12). In contrast, networks in the Texas Coastal Uplands aquifer system had consistently low percentages of samples classified as oxic-potentially young. The number of well networks in each of those aquifers, however, was relatively small compared to the Floridan aquifer system and Central glacial aquifers (fig. 12).

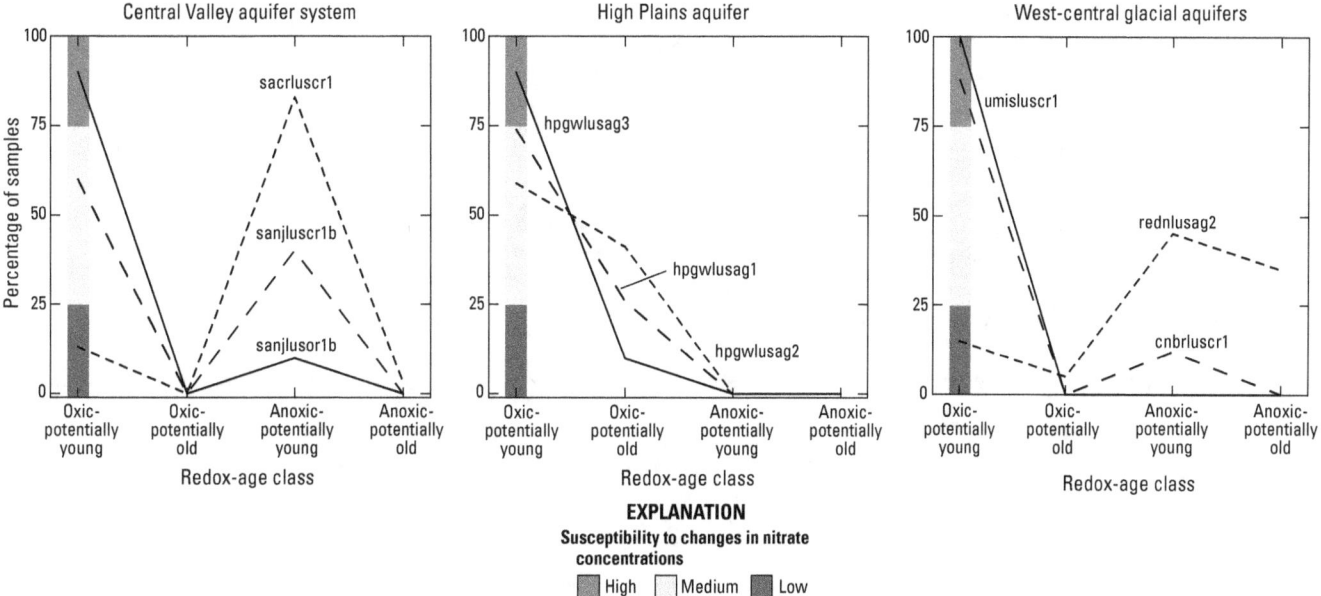

Figure 9.   Percentage of samples assigned to the four redox-age classes for networks of shallow monitoring wells in agricultural areas in the Central Valley aquifer system, High Plains aquifer, and the West-central glacial aquifers, and the susceptibility of the networks to changes in nitrate concentrations.

Figure 10.   Central locations of networks of domestic wells and the susceptibility of the networks to changes in nitrate concentrations.

Table 5. Redox-age classes for water samples collected from networks of domestic wells in the United States and the susceptibility of the networks to changes in nitrate concentrations (only networks with at least 10 wells are listed).

[usg, unconsolidated sand and gravel; gla, glacial sand and gravel; scs, semiconsolidated sand; san, sandstone; scr, sandstone and carbonate rock; car, carbonate rock; bav, basaltic and other volcanic rock; cry, crystalline rock; shading is used to differentiate between aquifer lithologies]

Aquifer number	Aquifer lithology	Aquifer name	Network name (number of wells)	Network identifier (see figure 10)	Redox-age class (percentage of samples)[5]				Susceptibility to changes in nitrate concentrations
					Oxic-potentially young	Oxic-potentially old	Anoxic-potentially young	Anoxic-potentially old	
1	usg	Basin and Range basin-fill aquifers	cazbsus3 (15)	d1	93	7	0	0	High
1	usg	Basin and Range basin-fill aquifers	cazbsus1b (17)	d2	82	0	18	0	High
1	usg	Basin and Range basin-fill aquifers	cazbsus2 (22)	d3	77	18	5	0	High
1	usg	Basin and Range basin-fill aquifers	grslsus1a (17)	d4	71	29	0	0	Medium
1	usg	Basin and Range basin-fill aquifers	grslsus1b (16)	d5	63	13	19	6	Medium
1	usg	Basin and Range basin-fill aquifers	nvbrsus3 (10)	d6	50	50	0	0	Medium
1	usg	Basin and Range basin-fill aquifers	cazbsus1a (21)	d7	38	57	0	5	Medium
3	usg	Central Valley aquifer system	sanjulusor1a (28)	d8	93	4	4	0	High
3	usg	Central Valley aquifer system	sanjulusor2a (26)	d9	92	0	8	0	High
3	usg	Central Valley aquifer system	sanjuluscr1a (24)	d10	83	0	13	4	High
3	usg	Central Valley aquifer system	sanjsus1 (32)	d11	81	6	6	6	High
3	usg	Central Valley aquifer system	sacrsus1 (26)	d12	58	15	23	4	Medium
4	usg	Columbia Plateau basin-fill aquifers	ccptlusag2a (12)[1]	d13	100	0	0	0	High
5	usg	High Plains aquifer	hpgwsus2 (20)	d14	90	0	10	0	High
5	usg	High Plains aquifer	hpgwsus1b (46)	d15	70	26	4	0	Medium
5	usg	High Plains aquifer	hpgwsus1a (74)	d16	46	53	1	0	Medium
5	usg	High Plains aquifer	hpgwsus1c (108)	d17	45	45	4	6	Medium
5	usg	High Plains aquifer	hpgwsus4 (30)2	d18	43	23	23	10	Medium
5	usg	High Plains aquifer	hpgwsus5 (27)	d19	41	37	11	11	Medium
7	usg	Northern Rocky Mountains Intermontaine Basins aquifer system	nroksus1 (29)	d20	55	28	14	3	Medium
7	usg	Northern Rocky Mountains Intermontaine Basins aquifer system	nroksus2 (28)	d21	54	39	7	0	Medium
8	usg	Rio Grande aquifer system	riogsus1 (24)	d22	4	21	4	71	Low
9	usg	Snake River Plain basin-fill aquifers	usnkluscr1 (25)	d23	64	0	32	4	Medium
10	usg	Surficial aquifer system	soflsus2 (30)	d24	13	10	50	27	Low
11	usg	Willamette Lowland aquifer system	willsus1 (66)	d25	52	5	32	12	Medium
12e	gla	Eastern glacial aquifers	connsus2 (28)	d26	57	4	32	7	Medium
12e	gla	Eastern glacial aquifers	delrsus3 (12)	d27	50	33	8	8	Medium
12e	gla	Eastern glacial aquifers	almnsus2 (30)	d28	50	0	43	7	Medium
12e	gla	Eastern glacial aquifers	hdsnsus3 (23)	d29	13	4	17	65	Low
12c	gla	Central glacial aquifers	uirbsus1 (27)	d30	67	0	33	0	Medium
12c	gla	Central glacial aquifers	lirbsus2 (28)	d31	57	0	29	14	Medium
12c	gla	Central glacial aquifers	wmicsus2 (25)	d32	48	12	20	20	Medium
12c	gla	Central glacial aquifers	uirbsus2 (23)	d33	48	0	52	0	Medium
12c	gla	Central glacial aquifers	miamsus1 (30)	d34	43	0	50	7	Medium
12c	gla	Central glacial aquifers	lerisus1 (30)	d35	30	20	23	27	Medium
12c	gla	Central glacial aquifers	lerispcg1 (21)	d36	19	10	57	14	Medium
12c	gla	Central glacial aquifers	lirbsus1 (26)	d37	0	0	23	77	Low
12wc	gla	West-central glacial aquifers	eiwasus2 (32)	d38	69	25	3	3	Low
12wc	gla	West-central glacial aquifers	hpgwsus4 (30)[2]	d39	43	23	23	10	Medium
12wc	gla	West-central glacial aquifers	rednsus2 (10)	d40	10	0	50	40	Low
12w	gla	Western glacial aquifers	pugtsus1 (29)	d41	72	10	10	7	Medium
12w	gla	Western glacial aquifers	cooksus1a (21)	d42	52	0	38	10	Medium

Table 5. Redox-age classes for water samples collected from networks of domestic wells in the United States and the susceptibility of the networks to changes in nitrate concentrations (only networks with at least 10 wells are listed).—Continued

[usg, unconsolidated sand and gravel; gla, glacial sand and gravel; scs, semiconsolidated sand; san, sandstone; scr, sandstone and carbonate rock; car, carbonate rock; bav, basaltic and other volcanic rock; cry, crystalline rock; shading is used to differentiate between aquifer lithologies]

Aquifer number	Aquifer lithology	Aquifer name	Network identifier (see figure 10)	Redox-age class (percentage of samples)[5]				Susceptibility to changes in nitrate concentrations
				Oxic-potentially young	Oxic-potentially old	Anoxic-potentially young	Anoxic-potentially old	
13	scs	Coastal Lowlands aquifer system	d43	74	11	16	0	Medium
13	scs	Coastal Lowlands aquifer system	d44	65	15	10	10	Medium
13	scs	Coastal Lowlands aquifer system	d45	27	64	0	9	Medium
13	scs	Coastal Lowlands aquifer system	d46	24	59	7	10	Low
14	scs	Mississippi Embayment aquifer system	d47	40	40	0	20	Medium
14	scs	Mississippi Embayment aquifer system	d48	18	6	24	53	Low
15	scs	Northern Atlantic Coastal Plain aquifer system	d49	81	0	6	13	High
15	scs	Northern Atlantic Coastal Plain aquifer system	d50	77	0	8	15	High
15	scs	Northern Atlantic Coastal Plain aquifer system	d51	77	0	23	0	High
16	scs	Southeastern Coastal Plain aquifer system	d52	72	0	22	6	Medium
16	scs	Southeastern Coastal Plain aquifer system	d53	41	23	18	18	Medium
17	scs	Texas Coastal Uplands aquifer system	d54	23	8	38	31	Low
17	scs	Texas Coastal Uplands aquifer system	d55	12	12	4	73	Low
18	san	Cambrian-Ordovician aquifer system	d56	54	8	33	4	Medium
18	san	Cambrian-Ordovician aquifer system	d57	44	8	24	24	Medium
18	san	Cambrian-Ordovician aquifer system	d58	41	0	27	32	Medium
19	san	Denver Basin aquifer system	d59	34	34	3	28	Medium
19	san	Denver Basin aquifer system	d60	12	15	19	54	Low
19	san	Denver Basin aquifer system	d61	0	10	0	90	Low
19	san	Denver Basin aquifer system	d62	0	20	20	60	Low
20	san	Early Mesozoic Basin aquifers	d63	88	0	8	4	High
20	san	Early Mesozoic Basin aquifers	d64	77	0	14	9	High
20	san	Early Mesozoic Basin aquifers	d65	70	10	5	15	Medium
21	san	Lower Tertiary aquifers	d66	33	5	29	33	Medium
22	san	Pennsylvanian aquifers	d67	48	3	45	3	Medium
22	san	Pennsylvanian aquifers	d68	13	0	81	6	Low
23	san	Valley and Ridge clastic-rock aquifers	d69	59	5	36	0	Medium
23	san	Valley and Ridge clastic-rock aquifers	d70	58	21	17	4	Medium
23	san	Valley and Ridge clastic-rock aquifers	d71	41	10	45	3	Medium
24	scr	Edwards-Trinity aquifer system	d72	85	7	7	0	High
24	scr	Edwards-Trinity aquifer system	d73	50	13	8	29	Medium
24	scr	Edwards-Trinity aquifer system	d74	43	21	14	21	Medium
29	car	Floridan aquifer system	d75	100	0	0	0	High
29	car	Floridan aquifer system	d76	67	20	13	0	Medium
29	car	Floridan aquifer system	d77	60	10	17	13	Medium
29	car	Floridan aquifer system	d78	12	8	42	38	Low
29	car	Floridan aquifer system	d79	10	28	24	38	Low
30	car	Ordovician aquifers	d80	19	19	52	10	Low
31	car	Ozark Plateaus aquifer system	d81	94	0	6	0	High
31	car	Ozark Plateaus aquifer system	d82	85	12	3	0	High

Network names (number of wells):

acadsus2 (19) — d43; acadsus1 (20) — d44; trinsus4 (11) — d45; trinsus3 (29) — d46; misesus4 (10) — d47; acadsus3 (17)[3] — d48; podlsus2 (16) — d49; linjsus2 (26) — d50; dlmvsus1 (13) — d51; moblsus3 (18) — d52; moblsus1 (22) — d53; acadsus3 (13)[3] — d54; sctxsus4 (26) — d55; umissus3 (24) — d56; umissus4 (25) — d57; wmicsus1 (22) — d58; spltsus2 (29) — d59; spltsus3 (26) — d60; spltsus4 (10) — d61; spltsus5 (10) — d62; delrsus1 (25) — d63; potosus2 (22) — d64; linjsus3 (20) — d65; yellsus2 (21) — d66; almnsus1 (29) — d67; kanasus1 (16) — d68; potolusag2 (22) — d69; delrsus2 (24) — d70; lsussus1 (29) — d71; sctxsus1 (27) — d72; sctxsus2 (24) — d73; trinsus1 (14) — d74; acfbsus1 (10) — d75; gaflsus2 (30) — d76; gaflsus3 (30) — d77; gaflsus4 (26) — d78; santsus2 (29) — d79; ltensus2 (21) — d80; ozrklusag1a (17) — d81; ozrksus2a (33) — d82

Table 5. Redox-age classes for water samples collected from networks of domestic wells in the United States and the susceptibility of the networks to changes in nitrate concentrations (only networks with at least 10 wells are listed).—Continued

[usg, unconsolidated sand and gravel; gla, glacial sand and gravel; scs, semiconsolidated sand; san, sandstone; scr, sandstone and carbonate rock; car, carbonate rock; bav, basaltic and other volcanic rock; cry, crystalline rock; shading is used to differentiate between aquifer lithologies]

Aquifer number	Aquifer lithology	Aquifer name	Network name (number of wells)	Network identifier (see figure 10)	Redox-age class (percentage of samples)[5]				Susceptibility to changes in nitrate concentrations
					Oxic-potentially young	Oxic-potentially old	Anoxic-potentially young	Anoxic-potentially old	
31	car	Ozark Plateaus aquifer system	ozrklusag2a (16)	d83	81	0	19	0	High
31	car	Ozark Plateaus aquifer system	ozrksus3a (16)	d84	56	19	13	13	Medium
32	car	Piedmont and Blue Ridge carbonate-rock aquifers	lsuslusag1 (29)	d85	83	0	17	0	High
33	car	Silurian–Devonian aquifers	eiwasus1 (25)	d86	16	4	20	60	Low
34	car	Valley and Ridge carbonate-rock aquifers	lsuslusag3 (29)	d87	100	0	0	0	High
34	car	Valley and Ridge carbonate-rock aquifers	lsuslusag2 (29)	d88	97	0	3	0	High
34	car	Valley and Ridge carbonate-rock aquifers	potolusag1 (32)	d89	94	0	6	0	High
34	car	Valley and Ridge carbonate-rock aquifers	utensus1 (18)	d90	67	11	22	0	Medium
35	bav	Columbia Plateau basaltic-rock aquifer system	ccptlusag1a (17)	d91	82	6	0	12	High
35	bav	Columbia Plateau basaltic-rock aquifer system	ccptlusag2a (16)[1]	d92	81	13	6	0	High
36	bav	Snake River Plain basaltic-rock aquifer system	usnkluscr4 (15)	d93	100	0	0	0	High
36	bav	Snake River Plain basaltic-rock aquifer system	usnkluscr2 (28)	d94	100	0	0	0	High
36	bav	Snake River Plain basaltic-rock aquifer system	usnkluscr3 (28)	d95	96	4	0	0	High
38	cry	New York and New England crystalline-rock aquifers	linjsus1 (25)	d96	88	8	4	0	High
38	cry	New York and New England crystalline-rock aquifers	necbsus2 (30)	d97	53	0	47	0	Medium
38	cry	New York and New England crystalline-rock aquifers	necbsus1 (28)	d98	46	0	54	0	Medium
38	cry	New York and New England crystalline-rock aquifers	connsus1 (27)	d99	44	33	4	19	Medium
39	cry	Piedmont and Blue Ridge crystalline-rock aquifers	lsussus2 (29)	d100	100	0	0	0	High
39	cry	Piedmont and Blue Ridge crystalline-rock aquifers	santsus3 (29)	d101	86	0	14	0	High
39	cry	Piedmont and Blue Ridge crystalline-rock aquifers	potosus1 (21)	d102	81	0	19	0	High
39	cry	Piedmont and Blue Ridge crystalline-rock aquifers	kanasus2 (19)	d103	74	5	16	5	Medium
39	cry	Piedmont and Blue Ridge crystalline-rock aquifers	albesus8 (48)	d104	67	10	21	2	Medium
--[4]	usg	Alluvial aquifers in the Colorado Rocky Mountains	ucolsus1 (23)	d105	78	13	9	0	High

[1]Network has wells in the Columbia Plateau basaltic-rock aquifer system and Columbia Plateau basin-fill aquifers.

[2]Network is part of the High Plains aquifer and West-central glacial aquifers.

[3]Network has wells in the Mississippi embayment and Texas coastal uplands aquifer systems.

[4]Network not in a principal aquifer.

[5]Redox-age percentages may not sum to 100 percent because of rounding.

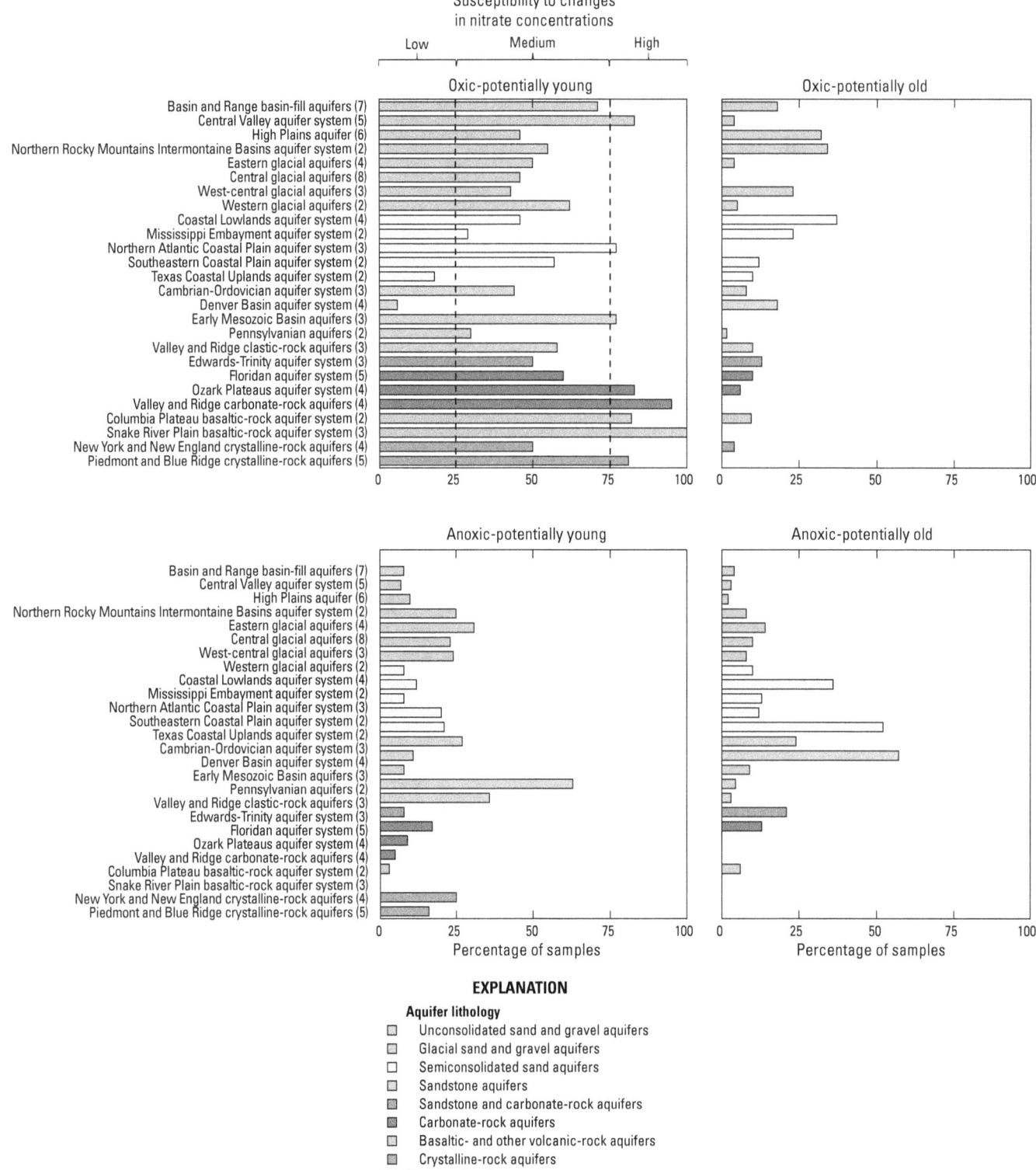

Figure 11.   Median percentage of samples assigned to the four redox-age classes for principal aquifers that have at least two networks of domestic wells, and the susceptibility of the aquifers to changes in nitrate concentrations.

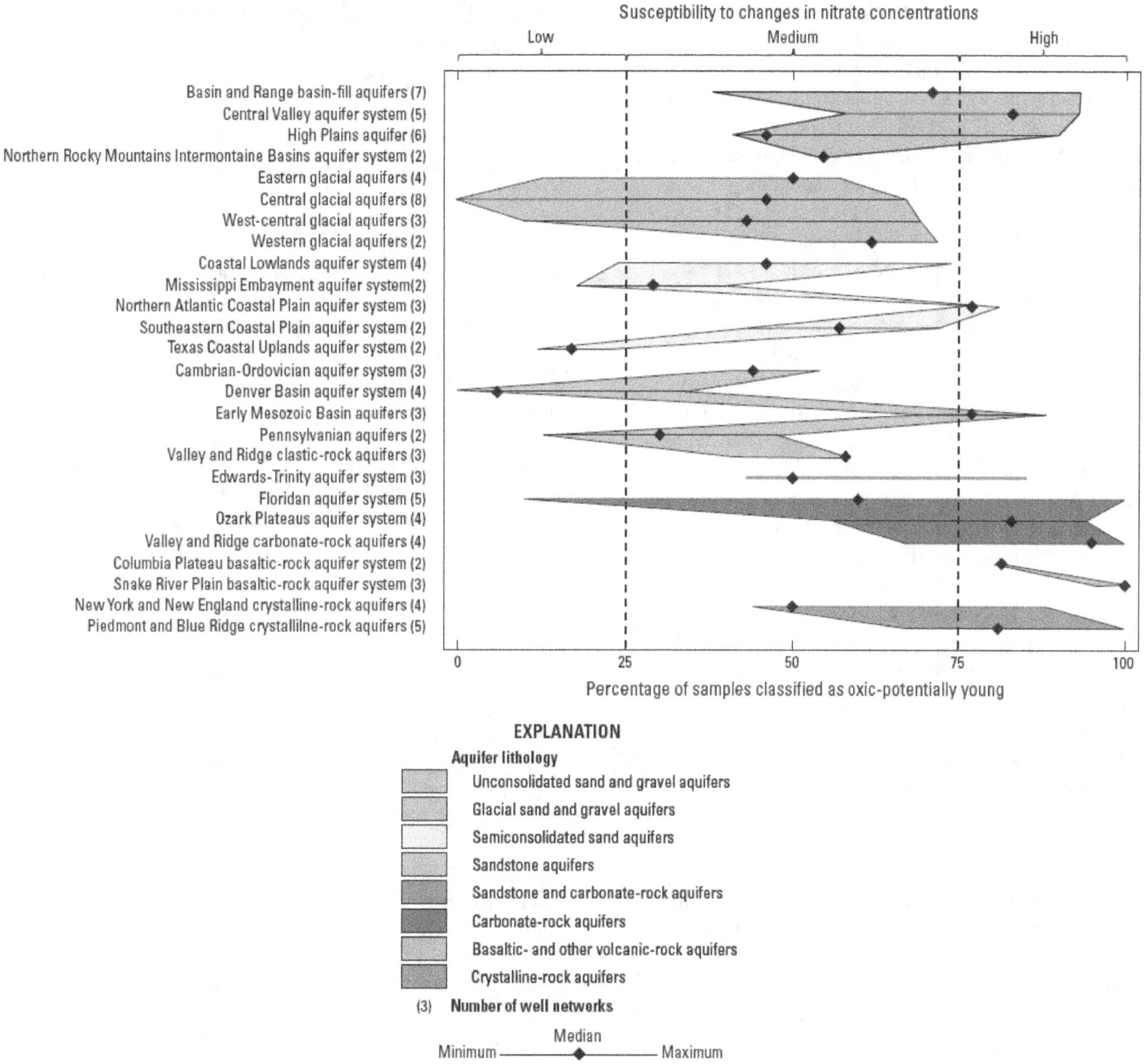

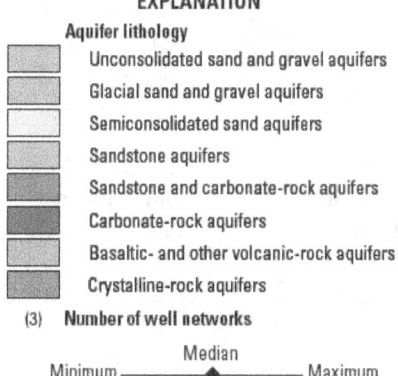

Figure 12. Variability in the percentage of samples classified as oxic-potentially young for principal aquifers that have at least two networks of domestic wells, and the susceptibility of the networks to changes in nitrate concentrations.

Principal-aquifer lithology groups with the largest percentage of networks considered to have a high susceptibility to changes in nitrate concentrations were the basaltic- and other volcanic-rock aquifer systems (100 percent of networks), carbonate-rock aquifers (50 percent), and crystalline-rock aquifers (44 percent) (table 5 and figs. 10 and 12). These three lithology groups include five of the six domestic-well networks with 100 percent of their samples classified as oxic-potentially young. The lithology groups with the smallest percentage of networks considered to have a high susceptibility to changes in nitrate concentrations were the glacial aquifers (0 percent of

networks) and sandstone aquifers (about 13 percent) (table 5 and fig. 10). These two lithology groups include the three well networks with 0 percent of their samples classified as oxic-potentially young (table 5).

There are important geologic differences between the aquifer lithology groups with high and low susceptibilities to changes in nitrate concentrations. The relatively large percentage of high-susceptibility networks in the basaltic- and other volcanic-rock aquifer systems, carbonate-rock aquifers, and crystalline-rock aquifers may indicate the importance of fractures and karst features in promoting the rapid movement of oxic-potentially young groundwater in those aquifers (Dubrovsky and others, 2010; McMahon and others, 2011). The relatively small percentage of high-susceptibility networks in the glacial and sandstone aquifers reflects geologic characteristics of those aquifers that support anoxic redox conditions (high electron donor content) and inhibit water movement (fine-grained confining layers).

Domestic-well networks in the eastern and western United States differed with respect to the percentage of samples assigned to certain redox-age classes. The 45 networks located in the western United States had a larger median percentage (13 percent) of samples classified as oxic-potentially old than the 60 networks located in the East (4 percent). Previous studies already noted the presence of oxic groundwater that was sometimes thousands of years old in organic carbon-poor unconsolidated sand and gravel aquifers in the western United States, particularly in the Central Valley and Rio Grande aquifer systems (Plummer and others, 2004; Jurgens and others, 2008), and the Basin and Range basin-fill and High Plains aquifers (Winograd and Robertson, 1982; McMahon and others, 2004). Those aquifers typically have low natural recharge rates and large, thick flow systems. In contrast, networks located in the eastern United States had a larger median percentage (20 percent) of samples classified as anoxic-potentially young than networks located in the West (7 percent). This is not surprising considering the generally shallower depths to water, higher natural recharge rates, and smaller, shallower flow systems in the eastern United States than in the West (Wolock, 2003; Reilly and others, 2008; McMahon and others, 2011). Oxic-potentially old and anoxic-potentially young conditions both reduce aquifer susceptibility to changes in nitrate concentrations, but for different climatic, geologic, and hydrologic reasons.

Fifteen of the domestic-well networks were approximately collocated with networks of shallow monitoring wells in agricultural areas, which provides the opportunity to compare the susceptibility to changes in nitrate concentrations at different depths in the same aquifer area. The median depth of the domestic wells was greater than the median depth of the monitoring wells for each pair of well networks. Overall, the median difference in depth between domestic and monitoring wells was 13 m. For 10 of the 15 pairs of networks, the monitoring-well networks had the higher percentage of samples classified as oxic-potentially young (fig. 13), indicating that susceptibility tended to be higher at the shallower depths of the monitoring wells. For 7 of the 15 pairs of nested networks, susceptibility was in fact higher in the monitoring wells than the domestic wells. Six of those seven pairs are in glacial aquifers (fig. 13). Only 3 of the 15 pairs of nested networks showed higher susceptibilities in the domestic wells than in the monitoring wells. For 5 of the 15 pairs of networks, susceptibilities were generally the same in both well types.

Susceptibility to Changes in Nitrate Concentrations in Parts of Aquifers that Provide Public Water Supplies

Redox-age classes were assigned to samples collected from 39 networks of public-supply wells (fig. 14 and table 6). Thirty-one percent of the networks were considered to have a high susceptibility to changes in nitrate concentrations and 26 percent of the networks were considered to have low susceptibilities (table 6). The percentage of high-susceptibility networks for public-supply wells was the same as for domestic wells, but the public-supply wells had a larger percentage of low-susceptibility networks than the domestic wells.

For principal aquifers that had at least 2 networks of public-supply wells, the median percentage of samples classified as oxic-potentially young ranged from about 7 to 87 percent (fig. 15), compared to about 57 to 96 percent for the shallow monitoring wells (fig. 8) and about 6 to 100 percent for domestic wells (fig. 11). For the parts of aquifers that provide public water supplies, the aquifers most susceptible to changes in nitrate concentrations were the Eastern glacial aquifers and the California Coastal Basin, Basin and Range basin-fill, and High Plains aquifers in the West (figs. 14 and 15). The least susceptible aquifer was the Cambrian-Ordovician aquifer system in the upper Midwest (figs. 14 and 15).

Susceptibility to changes in nitrate concentrations in the networks of public-supply wells appeared to be controlled in part by aquifer confinement and well depth, as was the case for several of the networks of monitoring and domestic wells. Low-susceptibility networks in the Cambrian-Ordovician aquifer system had relatively large well depths (median depths of 124 to 558 m) and large percentages of wells completed in confined parts of the aquifer (median values of 0 to 100 percent) compared to networks in the four aquifers with the highest susceptibilities. Well networks in the Eastern glacial aquifers, for example, had median well depths of 17 to 118 m and percentages of wells completed in confined parts of the aquifer that ranged from 0 to 24 percent.

Well networks sanasus2 and sanasus3, in the California Coastal Basin aquifers (table 6 and fig. 14), had medium susceptibilities even though the median well depths (232 to 294 m) were relatively large and 56 percent of the sanasus2 wells were completed in confined parts of the aquifer. Several factors probably contributed to the susceptibility of those two networks. One factor is long well screens. Median well-screen lengths in the networks accounted for 60 to 68 percent of the well depths. Long well screens could increase the chances of mixing shallow, young water and deep, old water (Landon and others, 2010b). Another possible factor is artificial recharge that occurs in parts of the California Coastal Basin aquifers in southern California that could increase aquifer susceptibility to changes in nitrate concentrations by increasing recharge rates (Hamlin and others, 2002; McMahon and others, 2011). Pumping rate also may affect susceptibility but data were not available to evaluate this factor.

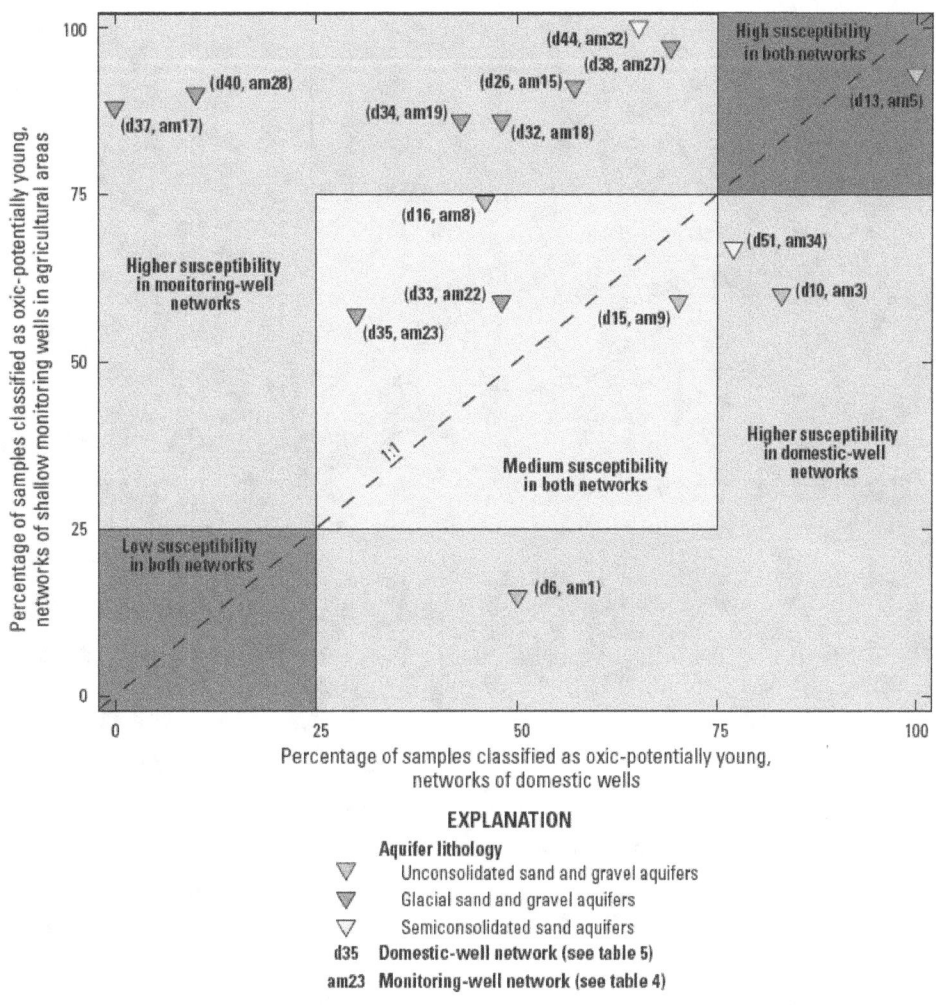

Figure 13. Percentage of samples classified as oxic-potentially young in collocated networks of shallow monitoring wells in agricultural areas and domestic wells, and the susceptibility of the networks to changes in nitrate concentrations.

Only four of the networks of public-supply wells were approximately collocated with networks of domestic wells. The median depth of the public-supply wells was greater than the median depth of the domestic wells for each pair of well networks. Overall, the median difference in depth between public-supply and domestic wells was 39 m, which is three times larger than the median difference in depth between the pairs of domestic- and monitoring-well networks. For three of the four pairs of networks, the public-supply wells had the higher percentage of samples classified as oxic-potentially young (fig. 16), indicating that susceptibility tended to be higher in the vicinity of public-supply wells than in the vicinity of domestic wells even though the public-supply wells had larger median well depths. Although the number of paired networks of public-supply and domestic wells was small, this finding is the opposite of what was observed for shallow monitoring wells and domestic wells (fig. 13). For one pair of public-supply (p24) and domestic-well networks (d45) (fig. 16), the percentage of samples classified as oxic-potentially young was higher for the

domestic wells. This may be due to the fact that only 50 percent of the domestic wells were completed in confined parts of the aquifer (Coastal Lowlands aquifer system) whereas 80 percent of the public supply wells were completed in confined parts of the aquifer. Bruce and Oelsner (2001) studied closely located pairs of domestic and public-supply wells in the High Plains aquifer and found a more frequent occurrence of pesticide compounds and tritium in water from the public-supply wells than in water from the domestic wells. They concluded that high rates of pumping in public-supply wells with long screens induced more rapid downward movement of young groundwater than did domestic wells, which had shorter screens and were less heavily pumped. Jurgens and others (2008) studied a long-screened public-supply well in the Central Valley aquifer system and also found that well construction and operation induced downward movement of young groundwater. The data in figure 16 are consistent with the idea that construction and operation characteristics of public-supply wells can enhance the downward movement of young groundwater (Landon and others, 2010b).

Figure 14.   Central locations of networks of public-supply wells and the susceptibility of the networks to changes in nitrate concentrations.

Table 6. Redox-age classes for water samples collected from networks of public-supply wells in the United States and the susceptibility of the networks to changes in nitrate concentrations (only networks with at least 10 wells are listed).

[usg, unconsolidated sand and gravel; gla, glacial sand and gravel; scs, semiconsolidated sand; san, sandstone; scr, sandstone and carbonate rock; car, carbonate rock; bav, basaltic and other volcanic rock; cry, crystalline rock; shading is used to differentiate between aquifer lithologies]

Aquifer number	Aquifer lithology	Aquifer name	Network name (number of wells)	Network identifier (see figure 14)	Redox-age class (percentage of samples)[3]				Susceptibility to changes in nitrate concentrations
					Oxic-potentially young	Oxic-potentially old	Anoxic-potentially young	Anoxic-potentially old	
1	usg	Basin and Range basin-fill aquifers	grslsus3 (30)	p1	83	10	0	7	High
1	usg	Basin and Range basin-fill aquifers	nvbrsus2 (25)	p2	60	32	0	8	Medium
1	usg	Basin and Range basin-fill aquifers	nvbrdwgs1 (14)	p3	57	36	0	7	Medium
2	usg	California Coastal Basin aquifers	sacrsus3 (11)	p4	82	0	9	9	High
2	usg	California Coastal Basin aquifers	sanasus1 (29)	p5	79	21	9	0	High
2	usg	California Coastal Basin aquifers	sacrsus4 (17)	p6	71	6	18	6	Medium
2	usg	California Coastal Basin aquifers	sanasus2 (18)	p7	67	6	11	17	Medium
2	usg	California Coastal Basin aquifers	sanasus3 (17)	p8	47	24	12	18	Medium
3	usg	Central Valley aquifer system	sanjdwgs1 (15)	p9	100	0	0	0	High
5	usg	High Plains aquifer	hpgwspcg7 (15)	p10	73	27	0	0	Medium
5	usg	High Plains aquifer	hpgwdwgs1 (15)[1]	p11	47	7	33	13	Medium
6	usg	Mississippi River Valley alluvial aquifer	misesus1 (10)	p12	10	0	80	10	Low
8	usg	Rio Grande aquifer system	riogdwgs1 (16)	p13	56	31	6	6	Medium
8	usg	Rio Grande aquifer system	riogtanc (23)	p14	39	39	0	22	Medium
9	usg	Snake River Plain basin-fill aquifers	usnksus3 (12)	p15	8	92	0	0	Low
10	usg	Surficial aquifer system	sofldwgs1 (15)	p16	40	0	60	0	Medium
12e	gla	Eastern glacial aquifers	linjdwgs1 (12)	p17	100	0	0	0	High
12e	gla	Eastern glacial aquifers	conmdwgs1 (15)	p18	87	0	13	0	High
12e	gla	Eastern glacial aquifers	necbsus3 (29)	p19	79	3	17	0	High
12c	gla	Central glacial aquifers	miamspcb1 (15)	p20	33	0	60	7	Medium
12c	gla	Central glacial aquifers	whmidwgs1 (13)	p21	23	0	54	23	Low
12wc	gla	West-central glacial aquifers	hpgwdwgs1 (15)[1]	p22	47	7	33	13	Medium
12wc	gla	West-central glacial aquifers	umisdwgs1 (15)[2]	p23	13	20	13	53	Low
13	scs	Coastal Lowlands aquifer system	trinsus4 (11)	p24	18	36	18	27	Low
14	scs	Mississippi Embayment aquifer system	misesus4 (10)	p25	40	50	10	0	Medium
14	scs	Mississippi Embayment aquifer system	misesus2 (30)	p26	20	0	40	40	Low
15	scs	Northern Atlantic Coastal Plain aquifer system	dlmvspcg10 (30)	p27	83	0	17	0	High
16	scs	Southeastern Coastal Plain aquifer system	santsus1 (20)	p28	95	0	0	5	High
18	san	Cambrian-Ordovician aquifer system	umisdwgs1 (15)[2]	p29	20	20	40	20	Low
18	san	Cambrian-Ordovician aquifer system	uirbsus3 (29)	p30	7	7	0	86	Low
18	san	Cambrian-Ordovician aquifer system	eiwasus3 (28)	p31	0	41	3	55	Low
19	san	Denver Basin aquifer system	spltdwgs1 (12)	p32	25	33	17	25	Medium
24	scr	Edwards-Trinity aquifer system	sctxsus3 (21)	p33	95	0	5	0	High
27	car	Biscayne aquifer	soflsus1 (23)	p34	13	0	87	0	Low
29	car	Floridan aquifer system	gafldwgs1 (14)	p35	64	7	14	14	Medium
29	car	Floridan aquifer system	gafldwgs2 (14)	p36	29	7	14	50	Medium
35	bav	Columbia Plateau basaltic-rock aquifer system	ccptsus1b (22)	p37	55	18	0	27	Medium
37	bav	Hawaiian volcanic-rock aquifers	oahusus1 (23)	p38	100	0	0	0	High
39	cry	Piedmont and Blue Ridge crystalline-rock aquifers	podldwgs1 (15)	p39	100	0	0	0	High

[1]Network is part of the High Plains aquifer and West-central glacial aquifers.

[2]Network has wells in the west-central glacial aquifers and Cambrian-Ordovician aquifer system.

[3]Redox-age percentages may not sum to 100 percent because of rounding.

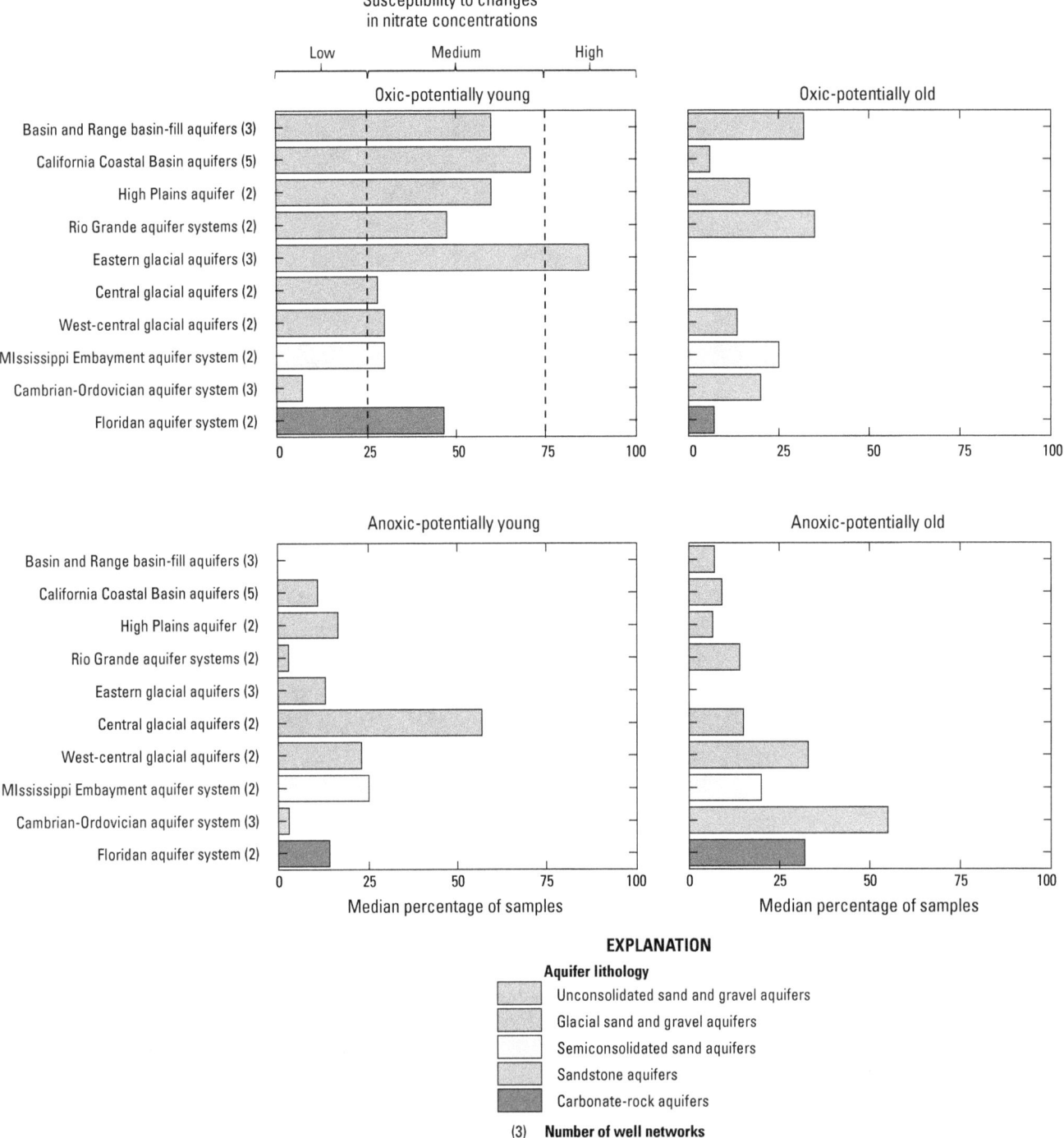

Figure 15.   Median percentage of samples assigned to the four redox-age classes for principal aquifers that have at least two networks of public-supply wells, and the susceptibility of the aquifers to changes in nitrate concentrations.

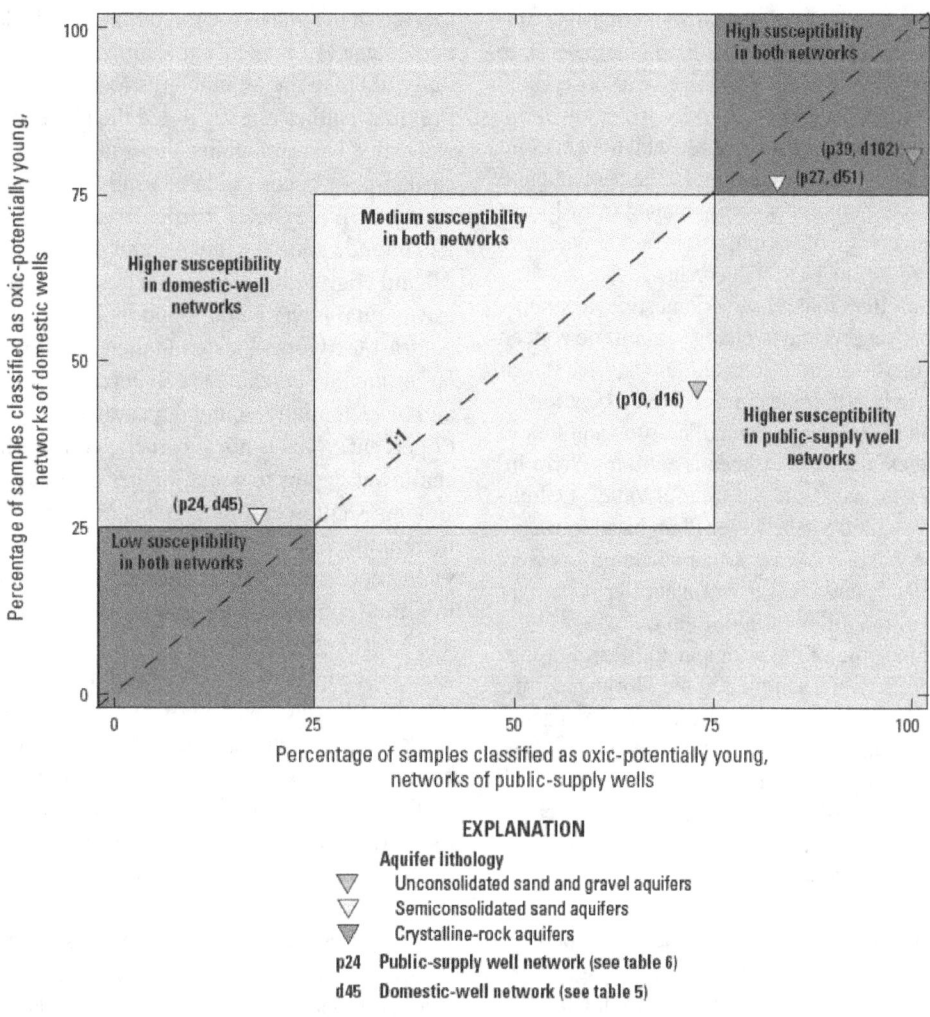

Figure 16. Percentage of samples classified as oxic-potentially young in collocated networks of public-supply and domestic wells, and the susceptibility of the networks to changes in nitrate concentrations.

Summary and Conclusions

The National Water-Quality Assessment (NAWQA) Program of the U.S. Geological Survey is using multiple approaches to measure and explain trends in concentrations of nitrate in principal aquifers of the United States. Near decadal sampling of selected well networks is providing information on where long-term changes in nitrate concentrations have occurred. Because those studies do not include all the NAWQA well networks, a determination has yet to be made as to what might be expected in networks from which time-series data have not been collected. Characterizing aquifer susceptibility to changes in nitrate concentrations on the basis of data collected from all the NAWQA well networks would be a step toward extrapolating findings from those studies to broader regions.

The purpose of this report is to characterize the susceptibility of selected principal aquifers of the United States to changes in nitrate concentrations on a basis of the redox-age classification scheme developed in this report. In this study, water samples collected from 6,593 wells in 39 principal

aquifers and 5 alluvial aquifers (collected from 1991 to 2010) were assigned to four redox-age classes on the basis of concentrations of dissolved oxygen and various indicators of groundwater age. The redox-age classes are oxic-potentially young, oxic-potentially old, anoxic-potentially young, and anoxic-potentially old. The redox-age assignments were then used to characterize the susceptibility of principal aquifers to changes in nitrate concentrations. Aquifer areas (as defined by well networks) in which at least 75 percent of the samples were classified as oxic-potentially young were considered to have a high susceptibility to changes in nitrate concentrations. Aquifer areas were considered to have a medium susceptibility if at least 25 percent and less than 75 percent of the samples were classified as oxic-potentially young. Aquifer areas were considered to have a low susceptibility if less than 25 percent of the samples were classified as oxic-potentially young.

For the parts of aquifers near the water table in agricultural areas, the aquifers most susceptible to changes in nitrate concentrations were the Columbia Plateau basin-fill aquifers, Eastern glacial aquifers, and the West-central glacial aquifers. None of the aquifers had a low susceptibility to changes in nitrate concentrations. Large intraaquifer redox-age variability was observed in most of the aquifers that had multiple networks of shallow monitoring wells in agricultural areas. For the three well networks in the Central Valley aquifer system, for example, the percentage of samples classified as oxic-potentially young ranged from 13 to 90 percent, and the percentage of samples classified as anoxic-potentially young ranged from 10 to 83 percent.

For the parts of aquifers that provide domestic water supplies, the aquifers most susceptible to changes in nitrate concentrations were the Northern Atlantic Coastal Plain aquifer system and the Early Mesozoic Basin, Valley and Ridge carbonate-rock, and Piedmont and Blue Ridge crystalline-rock aquifers in the eastern United States; the Ozark Plateaus aquifer system in parts of Missouri and Arkansas; and the Central Valley, Columbia Plateau basaltic-rock, and Snake River Plain basaltic-rock aquifer systems in the West. The least susceptible aquifers were the Texas Coastal Uplands and Denver Basin aquifer systems.

Relatively large intraaquifer variability in redox-age classes was observed in some of the principal aquifers. For the five networks of domestic wells sampled in the Floridan aquifer system, for example, the percentage of samples classified as oxic-potentially young ranged from 10 to 100 percent. Aquifer confinement probably is an important control on redox-age variability in the Floridan aquifer system. The Central glacial aquifers also showed large redox-age variability, which could be attributed to the diversity of depositional environments represented by well networks in those aquifers.

Principal-aquifer lithology groups with the largest percentage of domestic-well networks considered to have a high susceptibility to changes in nitrate concentrations were the basaltic- and other volcanic-rock aquifer systems, carbonate-rock aquifers, and crystalline-rock aquifers. These three lithology groups include five of the six domestic-well networks with 100 percent of their samples classified as oxic-potentially young. The lithology groups with the smallest percentage of networks considered to have a high susceptibility to changes in nitrate concentrations were the glacial aquifers and sandstone aquifers. These two lithology groups include the three well networks with 0 percent of their samples classified as oxic-potentially young. There are important geologic differences between the aquifer lithology groups with high and low susceptibilities to changes in nitrate concentrations. The relatively large percentage of high-susceptibililty networks in the basaltic- and other volcanic-rock aquifer systems, carbonate-rock aquifers, and crystalline-rock aquifers may indicate the importance of fractures and karst features in promoting the rapid movement of oxic-potentially young groundwater in those aquifers. The relatively small percentage of high-susceptibility networks in the glacial and sandstone aquifers reflects geologic characteristics of those aquifers that support anoxic redox conditions (high electron donor content) and inhibit water movement (fine-grained confining layers).

Domestic-well networks in the eastern and western United States differed with respect to the percentage of samples assigned to certain redox-age classes. The 45 networks located in the western United States had a larger median percentage (13 percent) of samples classified as oxic-potentially old than the 60 networks located in the East (4 percent). Previous studies already noted the presence of oxic groundwater that was sometimes thousands of years old in organic-carbon-poor unconsolidated sand and gravel aquifers in the western United States, particularly in the Central Valley and Rio Grande aquifer systems, and the Basin and Range basin-fill and High Plains aquifers. Those aquifers typically have low natural recharge rates and large, thick flow systems. In contrast, networks located in the eastern United States had a larger median percentage (20 percent) of samples classified as anoxic-potentially young than networks located in the West (7 percent). This is not surprising considering the generally shallower depths to water, higher natural recharge rates, and smaller, shallower flow systems in the eastern United States than in the West. Oxic-potentially old and anoxic-potentially young conditions both reduce aquifer susceptibility to changes in nitrate concentrations, but for different climatic, geologic, and hydrologic reasons.

Fifteen of the domestic-well networks were approximately collocated with networks of shallow monitoring wells in agricultural areas, which provided the opportunity to compare the susceptibility to changes in nitrate concentrations at different depths in the same aquifer area. The median depth of the domestic wells was greater than the median depth of the monitoring wells for each pair of well networks. For 10 of the 15 pairs of networks, the monitoring-well networks had the higher percentage of samples classified as oxic-potentially young, indicating that susceptibility tended to be higher at the shallower depths of the monitoring wells.

For the parts of aquifers that provide public water supplies, the aquifers most susceptible to changes in nitrate concentrations were the Eastern glacial aquifers and the California Coastal Basin, Basin and Range basin-fill, and High Plains aquifers in the West. The least susceptible aquifer was the Cambrian-Ordovician aquifer system in the upper Midwest.

Only four of the networks of public-supply wells were approximately collocated with networks of domestic wells. The median depth of the public-supply wells was greater than the median depth of the domestic wells for each pair of well networks. For three of the four pairs of networks, the public-supply wells had the higher percentage of samples classified as oxic-potentially young , indicating that susceptibility tended to be higher in the vicinity of public-supply wells than in the vicinity of domestic wells even though the public-supply wells had larger median well depths. Although the number of paired networks of public-supply and domestic wells was small, this finding is the opposite of what was observed for shallow monitoring wells and domestic wells. Previous studies found that high rates of pumping in public-supply wells with long screens induced more rapid downward movement of young groundwater than did domestic wells, which had shorter screens and were less heavily pumped. The data from this study are generally consistent with those findings.

Acknowledgments

Karen Burow and Matthew Landon provided helpful comments on earlier versions of this report.

References Cited

Böhlke, J.K., Wanty, R., Tuttle, M., Delin, G., and Landon, M., 2002, Denitrification in the recharge area and discharge area of a transient agricultural nitrate plume in a glacial outwash sand aquifer, Minnesota: Water Resources Research, v. 38, doi:10.1029/2001WR000663.

Böhlke, J.K., Verstraeten, I.M., and Kraemer, T.F., 2007, Effects of surface-water irrigation on sources, fluxes, and residence times of water, nitrate, and uranium in an alluvial aquifer: Applied Geochemistry, v. 22, p. 152–174.

Bruce, B.W., and Oelsner, G.P., 2001, Contrasting water quality from paired domestic/public supply wells, central High Plains: Journal of the American Water Resources Association, v. 37, p. 1389–1403.

Burow, K.R., Shelton, J.L., and Dubrovsky, N.M., 2008a, Regional nitrate and pesticide trends in groundwater in the eastern San Joaquin Valley, California: Journal of Environmental Quality, v. 37, p. S249–S263.

Burow, K.R., Jurgens, B.C., Kauffman, L.J., Phillips, S.P., Dalgish, B.A., and Shelton, J.L., 2008b, Simulations of ground-water flow and particle pathline analysis in the zone of contribution of a public-supply well in Modesto, eastern San Joaquin Valley, California: U.S. Geological Survey Scientific Investigations Report 2008–5035, 41 p.

Busenberg, E., and Plummer, L.N., 2000, Dating young ground water with sulfur hexafluoride—Natural and anthropogenic sources of sulfur hexafluoride: Water Resources Research, v. 36, p. 3011–3030.

Chapelle, F.H., McMahon, P.B., Dubrovsky, N.M., Fujii, R.F., Oaksford, E.T., and Vroblesky, D.A., 1995, Deducing the distribution of terminal electron-accepting processes in hydrologically diverse groundwater systems: Water Resources Research, v. 31, p. 359–371.

Clark, B.R., Landon, M.K., Kauffman, L.J., and Hornberger, G.Z., 2008, Simulations of ground-water flow, transport, age, and particle tracking near York, Nebraska, for a study of transport of anthropogenic and natural contaminants (TANC) to public supply wells: U.S. Geological Survey Scientific Investigations Report 2007–5068, 48 p.

Dubrovsky, N.M., Burow, K.R., Clark, G.M., Gronberg, J.M., Hamilton P.A., Hitt, K.J., Mueller, D.K., Munn, M.D., Nolan, B.T., Puckett, L.J., Rupert, M.G., Short, T.M., Spahr, N.E., Sprague, L.A., and Wilber, W.G., 2010, The quality of our Nation's waters—Nutrients in the Nation's streams and groundwater, 1992–2004: U.S. Geological Survey Circular 1350, 174 p.

Gilliom, R.J., Alley, W.M., and Gurtz, M.E., 1995, Design of the National Water-Quality Assessment Program: Occurrence and distribution of water-quality conditions: U.S. Geological Survey Circular 1112, 33 p.

Green, C.T., Puckett, L.J., Böhlke, J.K., Bekins, B.A., Phillips, S.P., Kauffman, L.J., Denver, J.M., and Johnson, H.M., 2008, Limited occurrence of denitrification in four shallow aquifers in agricultural areas of the United States: Journal of Environmental Quality, v. 37, p. 994–1009.

Green, C.T., Böhlke, J.K., Bekins, B.A., and Phillips, S.P., 2010, Mixing effects on apparent reaction rates and isotope fractionation during denitrification in a heterogeneous aquifer: Water Resources Research, v. 46, doi:10.1029/2009WR008903.

Gurdak, J.J., and Qi, S.L., 2006, Vulnerability of recently recharged groundwater in the High Plains aquifer to nitrate contamination: U.S. Geological Survey Scientific Investigations Report 2006–5050, 45 p.

Hamlin, S.N., Belitz, K., Kraja, S., and Dawson, B., 2002, Groundwater quality in the Santa Ana watershed, California, overview and data summary: U.S. Geological Survey Water-Resources Investigations Report 02–4243, 55 p.

Hinkle, S.R., Shapiro, S.D., Plummer, L.N., Busenberg, E., Widman, P.K., Casile, G.C., and Wayland, J.E., 2010, Estimates of tracer-based piston-flow ages of groundwater from selected sites—National Water-Quality Assessment Program, 1992–2005: U.S. Geological Survey Scientific Investigations Report 2010–5229, 90 p.

Jurgens, B.C., Burow, K.R., Dalgish, B.A., and Shelton, J.L., 2008, Hydrogeology, water chemistry, and factors affecting the transport of contaminants in the zone of contribution of a public-supply well in Modesto, eastern San Joaquin Valley, California: U.S. Geological Survey Scientific Investigations Report 2008–5156, 78 p.

Kaufman, S., and Libby, W.F., 1954, The natural distribution of tritium: Physics Review, v. 93, p. 1337–1344.

Kauffman, L.J., Baehr, A.L., Ayers, M.A., and Stackelberg, P.E., 2001, Effects of land use and travel time on the distribution of nitrate in the Kirkwood-Cohansey aquifer system in southern New Jersey: U.S. Geological Survey Water-Resources Investigations Report 01–4117, p. 49.

Kolpin, D.W., Goolsby, G.A., and Thurman, E.M., 1995, Pesticides in near-surface aquifers: An assessment using highly sensitive analytical methods and tritium: Journal of Environmental Quality, v. 24, p. 1125–1132.

Landon, M.K., Belitz, K., Jurgens, B.C., Kulongoski, J.T., and Johnson, T.D., 2010a, Status and understanding of groundwater quality in the Central–Eastside San Joaquin Basin, 2006—California GAMA Priority Basin project: U.S. Geological Survey Scientific Investigations Report 2009–5266, 97 p.

Landon, M.K., Jurgens, B.C., Katz, B.G., Eberts, S.M., Burow, K.R., and Crandal, C.A., 2010b, Depth-dependent sampling to identify short-circuit pathways to public-supply wells in multiple aquifer settings in the United States: Hydrogeology Journal, v. 18, p. 577–593.

Lindsey, B.D., and Rupert, M.G., 2012, Methods for evaluating temporal groundwater quality data and results of decadal-scale changes in chloride, dissolved solids, and nitrate concentrations in groundwater in the United States, 1988–2010: U.S. Geological Survey Scientific Investigations Report 2012–5049, 49 p.

Lucas, L.L., and Unterweger, M.P., 2000, Comprehensive review and critical evaluation of the half-life of tritium: Journal of Research of the National Institute of Standards and Technology, v. 105, p. 541–549.

Manning, A.H., Solomon, D.K., and Thiros, S.A., 2005, ^{3}H/^{3}He age data in assessing the susceptibility of wells to contamination: Ground Water, v. 43, p. 353–367.

McCulloch, A., 2003, Chloroform in the environment—Occurrence, sources, sinks and effects: Chemosphere, v. 50, p. 1291–1308.

McMahon, P.B., Böhlke, J.K., and Bruce, B.W., 1999, Denitrification in marine shales in northeastern Colorado: Water Resources Research, v. 35, p. 1629–1642.

McMahon, P.B., Böhlke, J.K., and Christenson, S.C., 2004, Geochemistry, radiocarbon ages, and paleorecharge conditions along a transect in the central High Plains aquifer, southwestern Kansas, USA: Applied Geochemistry, v. 19, p. 1655–1686.

McMahon, P.B., Dennehy, K.F., Bruce, B.W., Gurdak, J.J., and Qi, S.L., 2007, Water-quality assessment of the High Plains aquifer, 1999–2004: U.S. Geological Survey Professional Paper 1749, 136 p.

McMahon, P.B., and Chapelle, F.H., 2008, Redox processes and water quality of selected principal aquifer systems: Ground Water, v. 46, p. 259–271.

McMahon, P.B., Böhlke, J.K., Kauffman, L.J., Kipp, K.L., Landon, M.K., Crandall, C.A., Burow, K.R., and Brown, C.J., 2008a, Source and transport controls on the movement of nitrate to public supply wells in selected principal aquifers of the United States: Water Resources Research, v. 44, W04401, doi:10.1029/2007WR006252.

McMahon, P.B., Burow, K.R., Kauffman, L.J., Eberts, S.M., Böhlke, J.K., and Gurdak, J.J., 2008b, Simulated response of water quality in public supply wells to land use change: Water Resources Research, v. 44, W00A06, doi:10.1029/2007WR006731.

McMahon, P.B., Cowdery, T.K., Chapelle, F.H., and Jurgens, B.C., 2009, Redox conditions in selected principal aquifers of the United States: U.S. Geological Survey Fact Sheet 2009–3041, 6 p.

McMahon, P.B., Plummer, L.N., Böhlke, J.K., Shapiro, S.D., and Hinkle, S.R., 2011, A comparison of recharge rates in aquifers of the United States based on groundwater-age data: Hydrogeology Journal, v. 19, p. 779–800.

Michel, R.L., and Schroeder, R.A., 1994, Use of long-term tritium records from the Colorado River to determine timescales for hydrologic processes associated with irrigation in the Imperial Valley, California: Applied Geochemistry, v. 9, p. 387–401.

Mueller, D.K., and Helsel, D.R., 1996, Nutrients in the Nation's waters—Too much of a good thing?: U.S. Geological Survey Circular 1136, 24 p.

Nolan, B.T., and Hitt, K.J., 2003, Nutrients in shallow groundwaters beneath relatively undeveloped areas in the conterminous United States: U.S. Geological Survey Water-Resources Investigations Report 02–4289, 17 p.

Nolan, B.T., and Hitt, K.J., 2006, Vulnerability of shallow groundwater and drinking-water wells to nitrate in the United States: Environmental Science & Technology, v. 40, p. 7834–7840.

Pabich, W.J., Valiela, I., and Hemond, H.F., 2001, Relationship between DOC concentration and vadose zone thickness and depth below the water table in groundwater of Cape Cod, U.S.A.: Biogeochemistry, v. 55, p. 247–268.

Plummer, L.N., Michel, R.L., Thurman, E.M., and Glynn, P.D., 1993, Environmental tracers for age-dating young ground water, in Alley, W.M., ed., Regional groundwater quality, chap. 11: New York, Van Nostrand Reinhold, p. 255–294.

Plummer, L.N., Bohkle, J.K., and Busenberg, E., 2003, Approaches for ground-water dating, in Lindsey, B.D., Phillips, S.W., Donnelly, C.A., Speiran, G.K., Plummer, L.N., Böhlke, J.K., Focazio, M.J., Burton, W.C., and Busenberg, Eurybiades, Residence times and nitrate transport in ground water discharging to streams in the Chesapeake Bay Watershed: U.S. Geological Survey Water-Resources Investigations Report 03–4035, p. 12–24.

Plummer, L.N., Bexfield, L.M., Anderholm, S.K., Sanford, W.E., and Busenberg, E., 2004, Geochemical characterization of groundwater flow in the Santa Fe Group aquifer system, middle Rio Grande Basin, New Mexico: U.S. Geological Survey Water-Resources Investigations Report 03–4131, 395 p.

Plummer, L.N., Busenberg, E., Eberts, S.M., Bexfield, L.M., Brown, C.J., Fahlquist, L.S., Katz, B.G., and Landon, M.K., 2008, Low-level detections of halogenated volatile organic compounds in groundwater—Use in vulnerability assessments: Journal of Hydrologic Engineering, v. 13, p. 1049–1068.

Puckett, L.J., Tesoriero, A.J., Dubrovsky, N.M., 2011, Nitrogen contamination of surficial aquifers—A growing legacy: Environmental Science & Technology, v. 45, p. 839–844.

Reilly, T.E., Dennehy, K.F., Alley, W.M., and Cunningham, W.L., 2008, Groundwater availability in the United States: U.S. Geological Survey Circular 1323, 70 p.

Rosen, M.R., Voss, F.D., and Arufe, J.A., 2008, Evaluation of intra-annual variations in U.S. Geological Survey National Water-Quality Assessment groundwater quality data: Journal of Environmental Quality, v. 37, p. S199–S208.

Rupert, M.G., 1998, Probability of detecting atrazine/desethylatrazine and elevated concentrations of nitrate ($NO_2 + NO_3$–N) in groundwater in the Idaho part of the Upper Snake River Basin: U.S. Geological Survey Water-Resources Investigations Report 98–4203, 32 p.

Rupert, M.G., 2008, Decadal-scale changes of nitrate in groundwater of the United States, 1988–2004: Journal of Environmental Quality, v. 37, p. S240–S248.

Rupert, M.G., and Plummer, L.N., 2009, Groundwater quality, age, and probability of contamination, Eagle River watershed valley-fill aquifer, north-central Colorado, 2006–2007: U.S. Geological Survey Scientific Investigations Report 2009–5082, 59 p.

Shelton, J.L., Burow, K.R., Belitz, K., Dubrovsky, N.M., Land, M., and Gronberg, J.M., 2001, Low-level volatile organic compounds in active public supply wells as groundwater tracers in the Los Angeles physiographic basin, California, 2000: U.S. Geological Survey Water-Resources Investigations Report 01–4188, 35 p.

Tesoriero, A.J., and Puckett, L.J., 2011, O_2 reduction and denitrification rates in shallow aquifers: Water Resources Research, v. 47, doi:10.1029/2011WR010471.

Thatcher, L.L., 1962, The distribution of tritium fallout in precipitation over North America: Bulletin of the International Association of Scientific Hydrology, v. 7, p. 48–58.

U.S. Geological Survey, 2003, Principal aquifers, *in* National Atlas of the United States of America, 1 sheet, 1:5,000,000 scale, available at *http://www.nationalatlas.gov/wallmaps.html*.

U.S. Geological Survey, 2011, Publications of the National Water-Quality Assessment (NAWQA) Program: U.S. Geological Survey, accessed September 21, 2011, at *http://water.usgs.gov/nawqa/bib/*.

Weissmann, G.S., Zhang, Y., LaBolle, E.M., and Fogg, G.E., 2002, Dispersion of groundwater age in an alluvial aquifer system: Water Resources Research, v. 38, doi:10.1029/2001WR000907.

Winograd, I.J., and Robertson, F.N., 1982, Deep oxygenated groundwater—Anomaly or common occurrence?: Science, v. 216, p. 1227–1230.

Wolock, D.M., 2003, Estimated mean annual natural groundwater recharge in the conterminous United States: U.S. Geological Survey Open-File Report 03–311, digital data set.

Zogorski, J.S., Carter, J.M., Ivahnenko, T., Lapham, W.W., Moran, M.J., Rowe, B.L., Squillace, P.J., and Toccalino, P.L., 2006, Volatile organic compounds in the Nation's groundwater and drinking-water supply wells: U.S. Geological Survey Circular 1292, 101 p.

Appendix 1

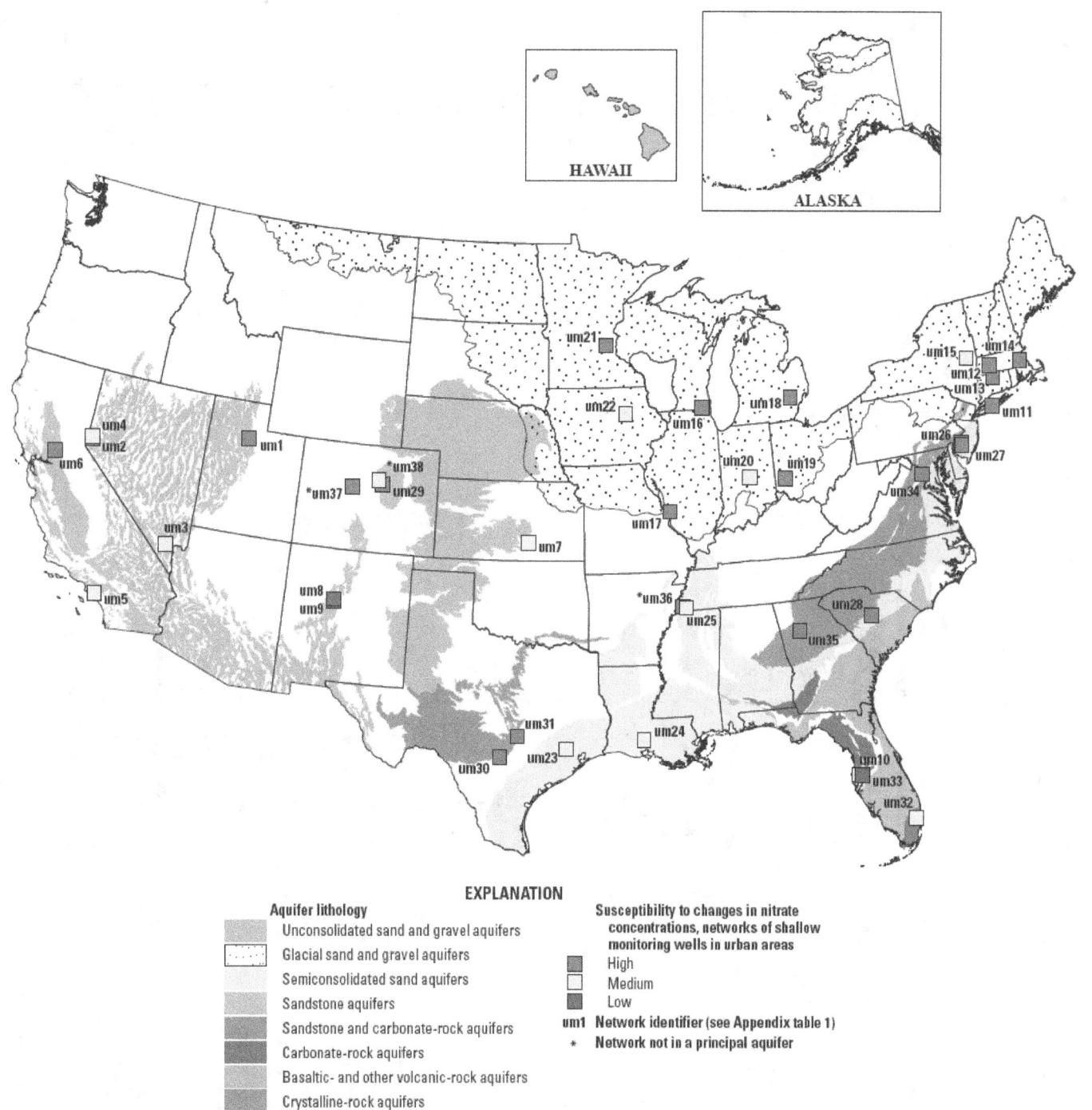

Figure 1–1.   Central locations of networks of shallow monitoring wells in urban areas and the susceptibility of the networks to changes in nitrate concentrations.

Table 1–1. Redox-age classes for water samples collected from networks of shallow monitoring wells in urban areas in the United States and the susceptibility of the networks to changes in nitrate concentrations (only networks with at least 10 wells are listed).

[usg, unconsolidated sand and gravel; gla, glacial sand and gravel; scs, semiconsolidated sand; san, sandstone; scr, sandstone and carbonate rock; car, carbonate rock; cry, crystalline rock; shading is used to differentiate between aquifer lithologies]

Aquifer number	Aquifer lithology	Aquifer name	Network name (number of wells)	Network identifier (see Appendix figure 1)	Redox-age class (percentage of samples)[2]				Susceptibility to changes in nitrate concentrations
					Oxic- potentially young	Oxic- potentially old	Anoxic- potentially young	Anoxic- potentially old	
1	usg	Basin and Range basin-fill aquifers	grsllusrc1 (29)	um1	100	0	0	0	High
1	usg	Basin and Range basin-fill aquifers	nvbrlusrc1 (16)	um2	81	13	6	0	High
1	usg	Basin and Range basin-fill aquifers	nvbrlusur1 (27)	um3	70	11	19	0	Medium
1	usg	Basin and Range basin-fill aquifers	nvbrlusur2 (20)	um4	35	10	50	5	Medium
2	usg	California Coastal Basin aquifers	sanalusrc1 (24)	um5	67	0	25	8	Medium
3	usg	Central Valley aquifer system	sacrlusrc1 (26)	um6	77	12	12	0	High
5	usg	High Plains aquifer	hpgwlusur1 (30)	um7	60	0	40	0	Medium
8	usg	Rio Grande aquifer system	rioglusur1 (24)	um8	17	4	42	38	Low
8	usg	Rio Grande aquifer system	rioglusrc1 (20)	um9	15	0	45	40	Low
10	usg	Surficial aquifer system	gafllusrc1a (12)	um10	58	0	42	0	Medium
12e	gla	Eastern glacial aquifers	linjlusrc2 (26)	um11	100	0	0	0	High
12e	gla	Eastern glacial aquifers	connlusur1 (39)	um12	90	8	0	3	High
12e	gla	Eastern glacial aquifers	connlusrc1 (27)	um13	81	0	19	0	High
12e	gla	Eastern glacial aquifers	necblusrc1 (29)	um14	76	0	24	0	High
12e	gla	Eastern glacial aquifers	hdsnlusur1 (16)	um15	50	19	19	13	Medium
12c	gla	Central glacial aquifers	uirblusrc1 (21)	um16	95	0	5	0	High
12c	gla	Central glacial aquifers	lirblusrc1 (25)	um17	80	16	4	0	High
12c	gla	Central glacial aquifers	lerilusrc1 (35)	um18	77	0	23	0	High
12c	gla	Central glacial aquifers	miamlusrc1 (24)	um19	75	0	25	0	Medium
12c	gla	Central glacial aquifers	whitlusur1a (25)	um20	56	0	40	4	High
12wc	gla	West-central glacial aquifers	umislusrc1 (32)	um21	88	0	13	0	Medium
12wc	gla	West-central glacial aquifers	eiwalusrc1 (29)	um22	55	0	45	0	Medium
13	scs	Coastal Lowlands aquifer system	trinlusrc1 (26)	um23	65	0	31	4	Medium
13	scs	Coastal Lowlands aquifer system	acadlusrc1 (25)	um24	48	16	28	8	Medium
14	scs	Mississippi Embayment aquifer system	miselusrc2 (10)	um25	50	30	20	0	Medium
15	scs	Northern Atlantic Coastal Plain aquifer system	linjlusur1 (20)	um26	100	0	0	0	High
15	scs	Northern Atlantic Coastal Plain aquifer system	linjlusrc1 (30)	um27	90	0	7	3	High
16	scs	Southeastern Coastal Plain aquifer system	santlusrc1 (30)	um28	90	7	3	0	High
19	san	Denver Basin aquifer system	spltlusrc2 (20)	um29	75	5	15	5	High
24	scr	Edwards-Trinity aquifer system	sctxlusrc1 (30)	um30	100	0	0	0	High
24	scr	Edwards-Trinity aquifer system	sctxlusrc2 (23)	um31	78	9	4	9	High
27	car	Biscayne aquifer	sofllusrc1a (30)	um32	60	7	30	3	Medium
29	car	Floridan aquifer system	gafllusrc1b (17)	um33	6	12	65	18	Low
39	cry	Piedmont and Blue Ridge crystalline-rock aquifers	podllusrc1 (30)	um34	90	0	10	0	High
39	cry	Piedmont and Blue Ridge crystalline-rock aquifers	acfblusur1 (15)	um35	80	13	7	0	High
--[1]	usg	Pleistocene Terrace deposits	miselusrc1 (26)	um36	77	4	15	4	High
--[1]	usg	Alluvial aquifers in the Colorado Rocky Mountains	ucollusrc1 (25)	um37	76	8	16	0	High
--[1]	usg	Denver Basin alluvial aquifers	spltlusrc1 (23)	um38	52	4	43	0	Medium

[1]Network not in a principal aquifer.

[2]Redox-age percentages may not sum to 100 percent because of rounding.

Publishing support provided by:
Denver Publishing Service Center, Denver, Colorado

For more information concerning this publication, contact:
Director, USGS Colorado Water Science Center
Box 25046, Mail Stop 415
Denver, CO 80225
(303) 236-4882

Or visit the Colorado Water Science Center Web site at:
http://co.water.usgs.gov/

This report is available at: *http://pubs.usgs.gov/sir/2012/5220*

www.ingramcontent.com/pod-product-compliance
Lightning Source LLC
Chambersburg PA
CBHW081621170526
45166CB00009B/3056